S0-BIC-130

Plan Reading for Home Builders

Plan Reading for Home Builders

J. Ralph Dalzell

SECOND EDITION

Revised by FREDERICK S. MERRITT
Consulting Engineer

McGRAW-HILL BOOK COMPANY
New York St. Louis San Francisco Düsseldorf Johannesburg
Kuala Lumpur London Mexico Montreal
New Delhi Panama Rio de Janeiro
Singapore Sydney Toronto

Library of Congress Cataloging in Publication Data

Dalzell, James Ralph, 1900–1970
Plan reading for home builders.

"An updating of Blueprint reading for home builders."
1. Architecture, Domestic—Designs and plans.
2. Architectural drawing. 3. Blue-prints.
I. Merritt, Frederick S., ed. II. Title.
TH431.D32 1972 692'.1 72-7068
ISBN 0-07-015221-7

34567890 HDBP·7654

The editors for this book were William G. Salo, Jr., and Lydia Maiorca, the designer was Naomi Auerbach, and its production was supervised by George E. Oechsner. It was set in Caledonia by Progressive Typographers. It was printed by Halliday Lithograph Corporation and bound by The Book Press.

Contents

Preface to Second Edition

In the many years since the first edition was published, the basic principles of reading construction plans have remained unchanged. The first edition was successful in teaching those principles to thousands of beginners. Hence, no major changes were necessary in producing this second edition.

Important changes, however, did take place in methods of reproducing drawings. Blueprints no longer are as widely used. Black and white prints are common. So a change in title of the book became advisable. In addition, numerous references to blueprints throughout the book had to be changed.

In the intervening years also, some of the symbols used on drawings have been changed. Newer materials are now more widely used in house construction. More appliances are being built into houses. And specifications are organized differently from the way they were at the time of the first edition. Though these changes have not affected basic principles, they required changes in text and illustrations to help students become familiar with the latest construction practices while learning to read construction plans. The second edition therefore represents principally an updating.

Frederick S. Merritt

Preface to First Edition

This book constitutes a great, interesting, and highly successful departure from the more or less precise styles employed in all previous books on blueprint reading. It is based on the experiences and knowledge the author has gained from seventeen previous and well-known building-trades textbooks, twenty years of architectural work, constant association with men in the building industry, many years as a teacher and as a writer of a nationally distributed column for homeowners which has reached and helped millions of people.

The book was planned, and is presented, so that it can serve the needs of students, men in the trades, homeowners, apprentices, real-estate people, appraisers, and anyone else interested in the plans with which new construction and remodeling work is carried on. No previous knowledge or experience of any kind is necessary in order to use the book to the best advantage.

Teachers will quickly find that this book accomplishes much more than merely assisting them in the teaching of blueprint reading and the fundamentals of other building-trade activity. Young mechanics and apprentices will find it of great help to them in all aspects of their advancement to journeyman status. Homeowners will find that it solves many of their problems and that it will help them with their improvement and maintenance plans.

Because blueprint reading is so much a matter of quick and proper visualization, each of the various chapters is introduced by easy-to-understand examples of visualization as it applies to the subject matter. The examples are all based on things with which the readers are already familiar, so that the gap between known and new concepts can be closed with ease and in an interesting manner.

The book is planned to carry on a constant and automatic review as each new item of subject matter is introduced. This presentation also helps to bridge the gap between known and new concepts. In other words, readers are taught to add new features to familiar mental pictures.

One of the many special features of the book is that three different types of houses are illustrated and explained throughout the chapters. A fourth type of house is employed for the questions at chapter ends. A fifth type is used in connection with the final chapter on blueprint reading. Thus, readers gain a varied experience as they learn to read blueprints.

The working drawings selected represent houses of the type which will endure, so far as popularity is concerned. Thus, this book should never be out of date in its application. The author has concentrated on the teaching of fundamentals which can be applied to reading blueprints representing all types of houses in any part of the country.

The section and detail view chapters were planned to serve as a primer so far as structural design is concerned. Thus, the book can serve as a steppingstone to design subjects and be invaluable in cases where builders are forced to design simple structural items.

Another feature of the book is the attention given to the relationships between plan, elevation, section, and detail views. Difficulties in learning are thus avoided and progress can be smooth and pleasant.

Still another feature has to do with the method of presentation. Each chapter is organized in the same manner so that readers will be traveling a familiar road all through the book.

As the various views and parts of working drawings are presented, they are explained in two ways. First, they are explained in a manner to make readers familiar with what they show. Finally, they are read as though reading working drawings.

In this book, none of the illustrations is shown merely as an example of miscellaneous structural detail. Instead, each of the illustrations, while showing all the general information necessary, is directly connected with working drawings or important terminology and thus becomes a vital teaching aid. The book contains only enough related information to associate blueprints with the architects who prepare them. No space is used for enrichment materials, because all the pages, in a book of this moderate size, are needed to explain fundamentals.

The book is presented in the descriptive language which the author has used so successfully in his previous textbooks. The text is easy to read, easy to understand, and easy to remember.

The author wishes to express his appreciation to the following sources for splendid cooperation:

The Small Homes Council
University of Illinois

The National Coal Association
Washington, D.C.

Samuel Paul, Architect
Jamaica, New York

Walter T. Anicka, Architect
Ann Arbor, Michigan

Campbell and Wong, Architects
San Francisco, California

J. Douglas Wilson, Formerly Supervisor
of Vocational Curricula
Los Angeles City Schools

R. Randolph Karch
State Department of Education
Harrisburg, Pennsylvania

Modern Homes Corporation
Dearborn, Michigan

Plan Reading for Home Builders

CHAPTER ONE

Introduction

A building-trades mechanic does not have to be concerned with the knowledge and skills which are needed to make construction drawings in order to read and interpret these drawings, or the various prints made from them. From the mechanic's viewpoint, it is only necessary to understand what such drawings are like and why they are required for all structural work.

All construction drawings are prepared using what can be called a special but quite simple language which any mechanic can learn with ease. The learning process has a definite beginning point and then progresses through various additional, easy-to-understand stages.

The purpose of this chapter is to explain the several fundamental aspects which constitute the starting point and to prepare readers for all subsequent learning stages.

CONSTRUCTION DRAWINGS

When a new house or the remodeling of an existing house is being planned, various types of construction drawings are required. There are several reasons why.

After an architect has discussed the building plans with his client, he has a fairly complete mental picture of the required house; and in his mind knows pretty well how it should look, how many rooms it is to have, what materials are to be used, the sizes of various structural parts, and many other details concerning it.

However, a mental picture at its best cannot be completely accurate or absolutely finished, because there are too many items to be concerned with and too many details which are impossible to design and correlate by means of mental pictures alone. It would be impossible for an architect to design and plan any building without drawing it on paper. Thus, an architect's mental picture must be converted to drawings where it can be further developed and completed.

Even if an architect could create accurate and complete mental pictures in his own mind, he would find it impossible to bring about exactly the same mental pictures in the minds of home-owners, carpenters, masons, and others interested in the building plans. Therefore, we see another reason why drawings are necessary. Once an architect completes a set of drawings, everyone concerned can read them and know exactly what the architect has in mind. In other words, the architect, the builders, and the owner will all have the same impressions and any possible misunderstanding can be avoided.

What Are Construction Drawings? From the foregoing, it is evident that construction drawings are picturelike representations which show how new or remodeled houses are to appear; what materials are to be employed; the number and sizes of rooms; sizes of all structural parts; types and sizes of windows and doors, closets, and storage spaces; plumbing; electrical work; and all other important details of construction. Such drawings show the builders exactly what to do in every phase of the structural work. In effect, the drawings constitute graphic instructions to builders. All necessary information is shown so that each and every detail of construction, from foundation to roof, is set forth in such a manner as to avoid any confusion or misunderstanding. In other words, construction drawings are step-by-step or how-to-do-it directions which are shown in picturelike form.

How Are Construction Drawings Made? First, the architect draws some more or less rough sketches. These help him to develop his own mental pictures and give the owner a chance to see the designs in picturelike form. Discussion between architect and owner continues until final agreement is reached.

Once the owner is satisfied with the rough sketches, the designer converts them into accurate and complete construction drawings. Naturally, the drawings have to be made in a size which can be easily handled. Therefore, they are miniature as well as picturelike reproductions of the structure they represent. We call them *scaled* drawings. If we take a picture of an existing structure, our picture may be about 3 by 4 inches, or smaller, in size. Yet, the picture is an exact reproduction of the structure on a very small scale.

Because construction drawings are so small, compared with the actual size of the structure they represent, a designer must employ a great many abbreviated and symbolic representations of the many materials and details necessary. Or, we might say, the designer must use a special kind of language to indicate the hundreds of items which he cannot actually picture, on such small size drawings. Thus, *symbols* are used to represent much of the information in connection with materials, windows, doors, bathroom fixtures, walls, foundations, floors, etc.

Several different types of construction drawings, as explained a little later in this chapter, are required in order to show all the needed information concerning new or remodeled structures. The drawings are prepared in such a way that each of them shows, in a standardized manner, certain required information. In other words, when we have learned how to use construction drawings for one structure, we will know how to use such drawings for any structure.

CONSTRUCTION PLANS

Similarly, engineers prepare engineering drawings to show how work they design should be constructed. Such drawings might indicate locations of electrical outlets or air-conditioning ducts in buildings, or the details of a bridge.

In any event, drawings show all the work required for construction of a structure. Hence, they are called *working drawings* or *construc-*

tion plans. Several types of drawings make up a complete *set* of working drawings. NOTE: Here we shall learn only the names of the various types of drawings. In later chapters we shall study them in detail.

Elevation Views This type of drawing shows what the exterior of a house or other structure is to look like. Four such drawings are required for an ordinary house. They contain a great deal of information and instructions for builders.

Plan View This type of drawing shows how a structure looks from above. For a house, a plan view shows the interior and indicates sizes, shapes, and arrangement of rooms, doors, windows, bathroom fixtures, kitchen equipment, etc. Plan and elevation views are two of the most important drawings we shall use.

Section View This type of drawing is used to show the interior construction of various structural parts.

Detail View This type of drawing is used to indicate required information about structural assembly, trim, and various special equipment. Such a view contains all of the information which cannot be shown by means of elevation, plan, or section views.

Mechanical Detail All detailed information in regard to electrical, heating, air-conditioning, and plumbing work is shown on this type of drawing.

Survey and Plot Plan A survey plan shows information concerning the site on which a structure is to be built. A plot plan shows where a structure is to be located on a site.

Perspective Drawing This type of drawing looks like a picture and is actually drawn by a designer to show an owner a "picture" of his structure before construction begins.

PRINTS

We can easily appreciate the fact that *one* set of working drawings as prepared by a designer could not serve the purpose of all people concerned. And it is also evident that a designer could not afford to prepare several sets of individually drawn plans. So, in order to provide many sets of working drawings, the original set is reproduced by a process somewhat similar to the developing and printing of photographs. Generally, the reproductions are black and white and are known as *prints.* However, any other color combination could be used just as well.

The terms *plans, working drawings,* and *prints* mean the same thing. From now on, we shall always speak of *drawings.*

HOW ARE DRAWINGS USED?

There are several most important uses for drawings. We shall consider such uses in terms of the various people who work with them.

Architects and Engineers They use sets of prints to show builders, estimators, and other interested people what their designs and instructions are. Often, an architect gives sets of drawings to a building contractor as a means of getting a bid on construction costs. A bid constitutes a proposal by a contractor to build a new structure or remodel an existing structure for a certain sum of money.

The Contractor As already indicated, a contractor uses sets of drawings to determine how much he wants to charge for the indicated structural work. He also needs drawings in getting approval from the city building department and as a means of getting building permits.

Builders We shall use the term *builders* to indicate the various mechanics, such as carpenters and masons. Builders use drawings as a guide for all the work they do. This fact will become clearer as subsequent chapters are read and studied.

Estimators Before the work on a new or remodeled structure is started, complete material and labor costs are generally assembled. Such work is done by experienced estimators who must use drawings for that purpose.

The Homeowner The owner uses sets of

drawings in planning furniture and decorations. He also needs drawings in dealing with finance organizations.

WRITTEN SPECIFICATIONS

Specifications are written instructions which must always accompany the drawings. In general, they include all phases of new or remodeling work which cannot be shown in the drawings. For example, specifications include instructions concerning all materials, methods of construction, standards of construction, and the manner of conducting all the work.

Specifications, as will be explained in a later chapter, supplement the drawings. In fact, the drawings are not complete without the specifications. Thus, all the specifications must be read and studied as carefully as the drawings.

VISUALIZATION

So far as the building trades are concerned, *visualization* is the process of looking at, or thinking about, a structural detail, or a symbol for such a detail, and being able to form a mental picture of how such a detail is constructed or assembled. For carpenters, masons, and all other building-trades mechanics, the ability to visualize is most important. Without that ability, reading of drawings could never be learned.

Earlier in this chapter, we discussed the necessity for drawings as a means of setting forth exactly the mental pictures the designer creates in designing or remodeling a structure. Thus, the ability to visualize (form mental pictures) is an absolute necessity for a designer. Conversely, it is just as necessary for carpenters and masons to visualize the drawings the designer creates.

If we read a story, we gradually form a mental picture of (visualize) the localities and people the author describes. We are able to see these localities and people in our mind's eye and to enjoy the story accordingly. Or, if we have seen various localities or people, we can recall them via our mental pictures (visualization) of them. We can describe the localities and people from our mental pictures.

In order to work as a building mechanic, we must be able to visualize all the parts of a structure from the representations and directions given in drawings and written specifications. For example, when drawings show a certain kind of wall, we must be able to visualize that kind of wall, both inside and out, and see all the associated parts. We cannot start carpentry or other work until we visualize the details we hope to construct.

The ability to visualize is not difficult to develop with practice and experience. For those readers who have not previously learned the process, the following examples indicate what is meant by visualization.

EXAMPLE 1. *Brick-veneer Wall.* The drawings and the specifications for a particular structure might indicate that brick-veneer exterior walls are required. In the drawings, the indication would be by symbols, which we shall learn in later chapters. In the written specifications, the wall would be described in terms of materials and various structural requirements, as we shall also learn in later chapters.

What should a mechanic be able to visualize when a brick wall is indicated? To answer that question, let us study the upper half of Figure 1, where the *A* part of the illustration shows only the exterior surface of the wall, and where the *B* part shows some of the interior structural items. In later chapters, we shall study such detail drawings much more carefully. Here, however, it is only necessary that we understand what visualization means. When a brick-veneer wall is mentioned, we should be able to visualize the numbered items, as shown in the *B* part of the illustration. We should learn to see these items without having to look at the illustration.

EXAMPLE 2. *Frame Wall with Wood Siding.* When such a wall is called for, we visualize the situation as previously explained. In the lower half of Figure 1, the *A* part of the illustration

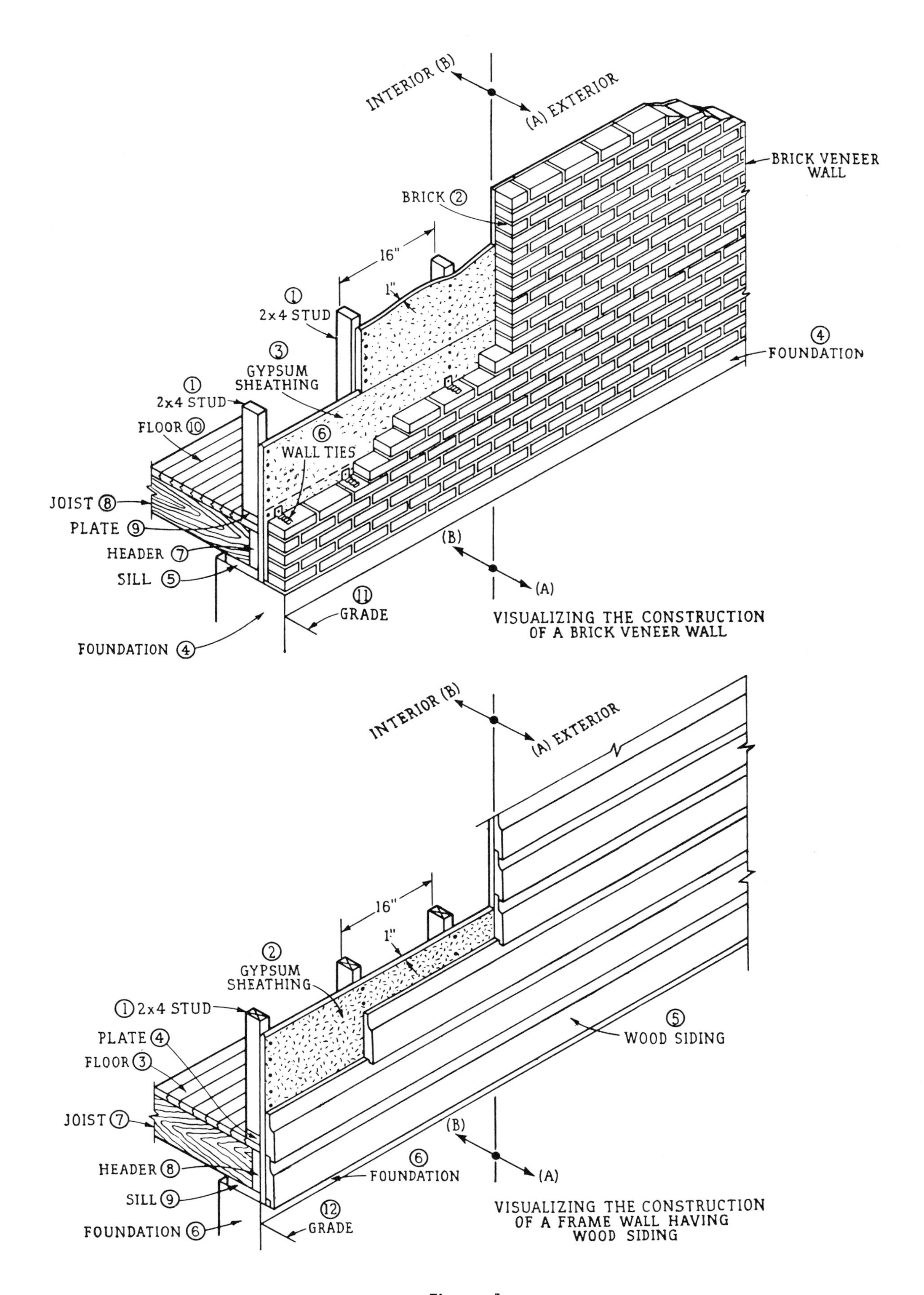
INTERIOR (B)
(A) EXTERIOR
BRICK VENEER WALL
BRICK ②
16"
1"
① 2x4 STUD
④ FOUNDATION
③ GYPSUM SHEATHING
① 2x4 STUD
FLOOR ⑩
⑥ WALL TIES
JOIST ⑧
PLATE ⑨
HEADER ⑦
SILL ⑤
(B)
(A)
⑪ GRADE
FOUNDATION ④
VISUALIZING THE CONSTRUCTION OF A BRICK VENEER WALL
INTERIOR (B)
(A) EXTERIOR
16"
1"
② GYPSUM SHEATHING
① 2x4 STUD
PLATE ④
FLOOR ③
⑤ WOOD SIDING
JOIST ⑦
(B)
HEADER ⑧
⑥ FOUNDATION
(A)
SILL ⑨
⑫ GRADE
FOUNDATION ⑥
VISUALIZING THE CONSTRUCTION OF A FRAME WALL HAVING WOOD SIDING

Figure 1

shows only the exterior surface (siding) of the wall, while the *B* part shows the numbered items we should be able to visualize whenever such a wall is mentioned.

Not all brick-veneer walls or frame walls with wood siding are exactly like those shown in Figure 1. Variations may occur from house to house. However, once we have learned to visualize, we can, in general, easily adapt our mental pictures to whatever conditions we encounter.

As previously mentioned, a designer uses many symbols (such as item 2 in Figure 1 for the brick-veneer wall and item 5 for the frame wall with wood siding) to designate certain materials and details of construction. In order to read drawings, we must learn to visualize all the symbols that architects and engineers use in construction plans.

HOW DOES A MECHANIC LEARN TO READ CONSTRUCTION PLANS?

Part of the answer to this question has been explained in the foregoing discussion about visualization. Other important steps to construction-plan reading are briefly mentioned in the following part of this chapter and explained fully in subsequent chapters.

Types of Drawings Certain information is always shown on specific drawings. For example, the brick-veneer and wood-siding symbols shown in Figure 1 are always indicated on elevation views. Thus, we must learn what information certain types of drawing contain.

Symbols We know that architects and engineers use many symbols in drawings. Some symbols are used only on elevation views. Others are used only on plan views. We must learn what each symbol means as it is employed on the various types of drawings.

Architectural and Engineering Terms Each part of a structure has a definite name. These names must be learned.

Abbreviations A great many abbreviations are employed in drawings. Their use avoids confusion and makes drawings much easier to read. Such abbreviations must be kept in mind.

Scaling Because drawings show all structures on a small scale, we must learn the various scales, and how to use them. In other words, we must learn to visualize drawings drawn to scales which are only fractional parts of full size. We must be able to look at drawings which are small enough to carry in our pockets and see in them the full-sized structures they represent.

Dimensions All shapes and sizes are represented by different kinds of dimensions on drawings. Therefore, we must become thoroughly acquainted with all dimensioning procedures.

Lines Many types of lines are used in drawings. Some of them indicate exterior edges. Others indicate imaginary cuttings to show detail interiors. Lines, like symbols, have important meanings.

There are many more aspects of construction-plan reading which must also be learned. This, however, is not difficult because, once we learn to visualize, each mark on a drawing means something to us, just as the letters of the alphabet mean something when we are reading a book or a newspaper. The following chapters contain complete explanations of everything we need to know in order to read construction plans easily and accurately.

WHY SHOULD MECHANICS LEARN TO READ DRAWINGS IN TERMS OF ALL TRADES?

When a structure is in the process of erection, all builders must work as a team. Unless they do work in such a spirit of cooperation, delays and costly confusion are bound to occur. For example, suppose that a reinforced-concrete slab is necessary for the first floor of a house. First, carpenters go about constructing the forms. At about the same time, plumbers, electricians, and steam fitters must make provisions for the conduits and pipes which will go in or through the concrete. Finally, the masons place the concrete. From this example, it can be seen that if any one of the trades represented should fail to cooperate with the others, costly delays and expensive mistakes could easily occur.

Cooperation between trades is not possible unless all of the men in the trades can read drawings in such a manner that they know what is required of the other trades as well as what is required of their own. This is especially true in regard to carpenters because they must guide and prepare the way for all other trades. Therefore, as we study and learn the process of reading drawings, we should learn to read the symbols and directions for all trades equally as well as our own.

QUESTIONS AND ANSWERS

Suggestions As a means of checking or testing the knowledge you have gained from your study of Chapter 1, try answering the following questions. After you answer each of the questions, read the answer shown to see how nearly right you were. Answering the questions is a good way to fix knowledge in your mind.

Question 1 Where would you look, among a set of drawings and written specifications, to find the material required for the outside surface of an exterior wall?
Answer 1 On the elevation views and in the written specifications.

Question 2 Why is it absolutely necessary for an architect to make drawings?
Answer 2 So that he can accurately and completely develop the plans and so that other interested people will understand exactly what the architect has in mind.

Question 3 Where would you look for information concerning the construction of details?
Answer 3 Among the detail drawings.

Question 4 How can misunderstandings between architects, owners, and builders be avoided?
Answer 4 Misunderstandings are prevented when architects make accurate and complete drawings. From such drawings all interested people should get the same impressions.

Question 5 Why should a carpenter be able to read the symbols and directions on drawings relating to all other trades?
Answer 5 So that he can cooperate with all other trades and guide the work.

Question 6 How do architects show most of the needed information on drawings?
Answer 6 By the use of symbols.

Question 7 Do all symbols mean exactly the same on all types of drawings?
Answer 7 No. Many of the symbols are used on specific types of drawings. For example, some of them are only used on plan views.

Question 8 What two types of architectural drawings are the most important?
Answer 8 Elevation and plan views.

Question 9 How can builders tell how a structure should be positioned on a site?
Answer 9 By referring to the plot plan.

Question 10 What is meant by visualization?
Answer 10 Visualization is the ability to form mental pictures of the various symbols on a drawing. For example, if a certain type of exterior wall is required, mechanics should be able to form a mental picture of that wall, both inside and out, which includes all of the structural items and materials.

Question 11 Is an accurate and complete set of drawings all one needs to learn everything about a proposed structure?
Answer 11 No. The written specifications are also very necessary because they supply information which cannot be shown on the drawings.

Question 12 Why is it that mental pictures cannot be depended upon for accuracy?
Answer 12 Because all the details cannot be correlated or worked out by mental pictures. There are more such details than the mind can cope with.

CHAPTER TWO

Scaling and Dimensioning

As explained in Chapter 1, construction plans are picturelike drawings which show all of the structural parts of a structure in such a manner that the designer can indicate just what he wants and the builders can understand exactly what they are to do. Also in Chapter 1, we discussed the fact that working drawings must be a great deal smaller than the structures they represent. When designers prepare drawings in greatly reduced sizes, they must scale their drawings in exact proportion to the true size of whatever they are designing. For this purpose they use many *scales*. Sizes of all structural parts are indicated by *dimensions*. As with scaling, dimensions must be shown in a specific manner, so that they will mean exactly the same to everyone reading the construction plans.

The purpose of this chapter is to explain scales, how drawings are made to scale, how dimensions are created and interpreted, how to find sizes when dimensions are not on the drawings, and other related information which will be of help to readers who want to learn to read construction plans.

SCALES

The process of drawing large objects, such as the parts of a structure, to a proportionate size which can be contained on handy-to-use sheets of paper is called *drawing to scale*. The small drawings must be in exact proportion to the actual size of the structure they represent. For example, for most buildings, the scale used makes the drawings 1/48 the original size. Or, in other words, instead of drawing something 1 foot long, we draw it 1/48 of 1 foot or ¼ inch long. Thus, the finished drawing looks like the full-sized object (is in exact proportion) but is only 1/48 the size.

The Architectural Scale The *A* part of Figure 2 shows part of what is known as a three-sided

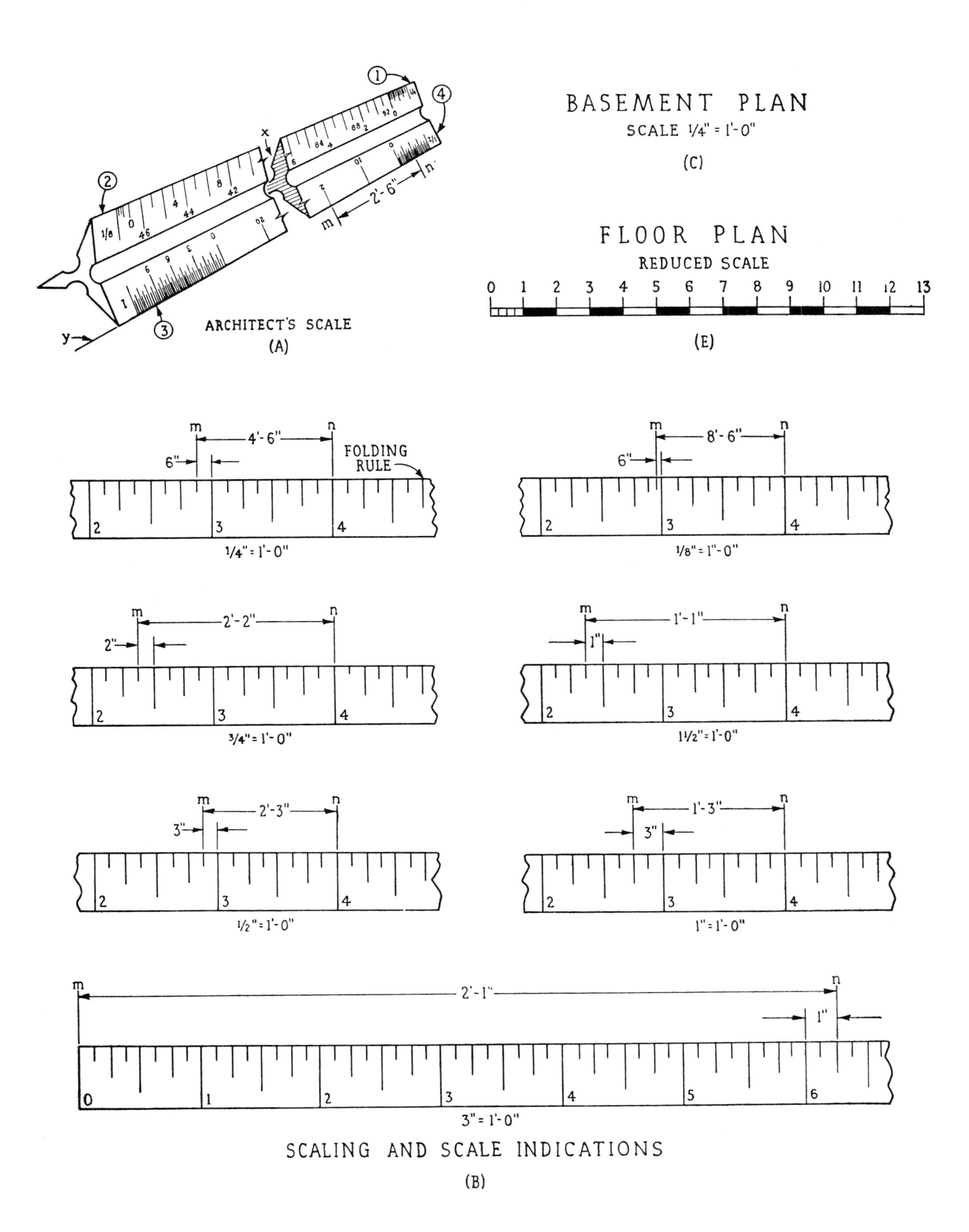

ARCHITECT'S SCALE
(A)
BASEMENT PLAN
SCALE 1/4" = 1'-0"
(C)
FLOOR PLAN
REDUCED SCALE
0 1 2 3 4 5 6 7 8 9 10 11 12 13
(E)
2'-6"
FOLDING RULE
4'-6"
6"
1/4" = 1'-0"
8'-6"
6"
1/8" = 1'-0"
2'-2"
2"
3/4" = 1'-0"
1'-1"
1"
1 1/2" = 1'-0"
2'-3"
3"
1/2" = 1'-0"
1'-3"
3"
1" = 1'-0"
2'-1"
1"
3" = 1'-0"
SCALING AND SCALE INDICATIONS
(B)

Figure 2

architects' or *architectural scale.* NOTE: The cut portion shown at *x* indicates that not all the scale is shown. Such a practice is often followed when designers want to indicate how an object appears but do not want to use enough space to show the full length. In fact, the scale is 12 inches long.

The three-sided scale is somewhat similar to an ordinary ruler except that, instead of having just one edge where inches, quarter inches, etc., are marked off, there are six edges, each of which shows two different scales.

The scale marked 1 is called the quarter-inch or ¼″ = 1′ 0″ scale. In 12 inches, there are 48 quarter inches. When using this scale, architects substitute ¼ inch for a foot and thus make their drawings to the 1/48 size mentioned previously. At the end of this scale one of the quarter-inch divisions is divided into 2, 4, and 12 smaller divisions. Each of the 12 small divisions represents 1 inch on the ¼″ = 1′ 0″ scale. Each of the 4 larger divisions represents 3 inches. Each of the 2 largest divisions represents 6 inches. EXAMPLE: Suppose that a designer wanted to draw a line or indicate a distance which is 12′ 6″ long, on a drawing where the ¼″ = 1′ 0″ scale is being used. The line he draws must be 12 quarter inches plus one of the two largest divisions at the end of the scale.

The scale marked 2 is called the eighth-inch or ⅛″ = 1′ 0″ scale. In 12 actual inches (the length of the scale) there are 96 eighth inches. When using this scale, architects substitute ⅛ inch for a foot and thus make their drawings 1/96 actual size. At the end of the scale, one of the eighth-inch divisions is divided into small divisions. The two largest of the small divisions each represent 6 inches. The smallest divisions each represent 2 inches. On such a small scale, any distance less than 2 inches must be estimated.

The scale marked 3 is called the 1-inch or 1″ = 1′ 0″ scale. When using this scale, architects substitute 1 inch for a foot and thus make their drawings 1/12 actual size. At the end of the scale, 1 inch is divided, first, into 4 large divisions each of which represents 3 inches. Each of the next smaller divisions represents an inch. The two smallest divisions represent half and quarter inches.

The scale marked 4 is called the half-inch or ½″ = 1′ 0″ scale. By the use of this scale, drawings can be made which are 1/24 actual size.

The other edges, which are not visible in the *A* part of Figure 2, contain the 3/16″ = 1′ 0″, the 3/32″ = 1′ 0″, the 1½″ = 1′ 0″, the ⅜″ = 1′ 0″, the ¾″ = 1′ 0″, the 3″ = 1′ 0″, and the 1/16″ = 1′ 0″ scales.

When Are Various Scales Used? The selection of scale depends to a great extent upon the size of the structure or structural parts being represented. For an ordinary structure, the elevation and plan view drawings are made to the ¼″ = 1′ 0″ scale. Detail drawings, where types of construction and materials are shown, are often drawn to the ½″ = 1′ 0″ or ¾″ = 1′ 0″ scale. If detail parts are very small, and an easy-to-read drawing is required, the 3″ = 1′ 0″ scale is often used. The ⅛″ = 1′ 0″ and the 1/16″ = 1′ 0″ scales, being very small, are often used for exceptionally large elevation views or for plot and survey plans. Use of such small scales keeps the overall size of drawings within reasonable limits. There are no special rules governing the scales to use and a designer generally uses his own judgment.

Scale Notation Note the *C* part of Figure 2. Each of the several drawings composing a set of construction plans has a title printed on it. For example, one of the drawings might be called the Basement Plan. Most designers indicate the scale used for each such drawing, as shown in the illustration.

DRAWING TO SCALE

As can be seen from the *A* part of Figure 2, the three-sided scale is so constructed that any one of the edges can be placed against drawing paper or a print. In the illustration, this is indicated by the edge marked y. Thus, we can see how easy it is to mark off scaled lengths and to measure unknown distances on drawings. EXAMPLE: Suppose we want to draw two lines, such as m and n, 2′ 6″ apart according to the ½″ = 1′ 0″ scale. To do that we put the edge of that scale on the paper and mark dots under the 2-foot and the 6-inch mark or division. Or, if the two lines appear on a print and the distance is not indicated, we can place the scale as shown in the A part of the illustration and find the distance to be 2′ 6″.

Use of the Folding Rule Carpenters and other mechanics are probaby more familiar with folding rules than they are with architects' scales. Therefore, we can discuss drawing to scale by the use of a representation of that type of rule. The B part of Figure 2 shows several examples where various scales are indicated. We can again assume that lines m and n have to be drawn certain distances apart and on certain scales. Or, we can assume that we want to find the distance between them on the scales shown. This is easy to do with a folding rule. The portion of the folding rule above the ¼″ = 1′ 0″ notation shows that the lines are 4′ 6″ apart, and that 4 quarter inches plus ⅛ inch represent 4′ 6″ according to that scale.

In like manner, portions of a folding rule, as shown in the illustration, indicate other distances according to the scales used. All this can be readily visualized by practice with an actual rule and a drawing.

DIMENSIONS

Aside from the picturelike and material representations on drawings, dimensions are the most important aspect of them. In other words, after visualizing a proposed new or remodeled structure from the viewpoint of style, shape, number of rooms, required materials, windows, doors, etc., builders want to know the sizes of all items. For example, they want to know the size and ceiling height of each room, the foundation depth and thickness, wall thicknesses, and hundreds of other size stipulations. Estimators are also highly interested in sizes because a large proportion of the costs they calculate concern the sizes of various materials.

All size or space stipulations on drawings are generally indicated (except as noted and explained in a following section of this chapter) by a system of lines, arrows, and figures which are known as *dimensions*. In all cases, dimensions are given in actual or full sizes or distances, regardless of the fact that construction plans show various drawings on a small scale.

The A part of Figure 3 shows an enclosure having two windows. We shall learn more about such drawings in a later chapter. Here, we are only interested in how dimensions are shown and what they indicate. This enclosure, as indicated by the scale notation, was drawn to the ¼″ = 1′ 0″ scale. The different dimensions can be checked using a three-sided scale or a folding rule. For example, the windows are 2′ 0″ wide. Therefore, they were drawn 2 quarter inches wide. The wall is 9 inches thick, so it was drawn equal to 9 of the small divisions, at the end of the scale, or 3/16 inch wide, as shown on a folding rule. The width of the enclosure is 12′ 6″, so it was drawn 12 quarter inches plus 6 of the small spaces on the three-sided scale, or ⅛ inch on the folding rule.

The 14′ 6″ and 12′ 6″ dimensions are known as the overall dimensions because they indicate the full exterior size of the enclosure. In both cases, the limits of the dimensions are indicated by the arrows. The actual sizes, in figures, are shown near the dimension lines. Extension lines are used to indicate what the arrows point to or where the dimension ends. For example, the extension lines indicate that the 14′ 6″ dimension line is the exterior length of side ac of the enclosure. In like manner, extension lines indicate that the 12′ 6″ dimension line is the exterior length of side cd of the enclosure.

In addition to the overall dimensions, the exact locations of the windows and the wall

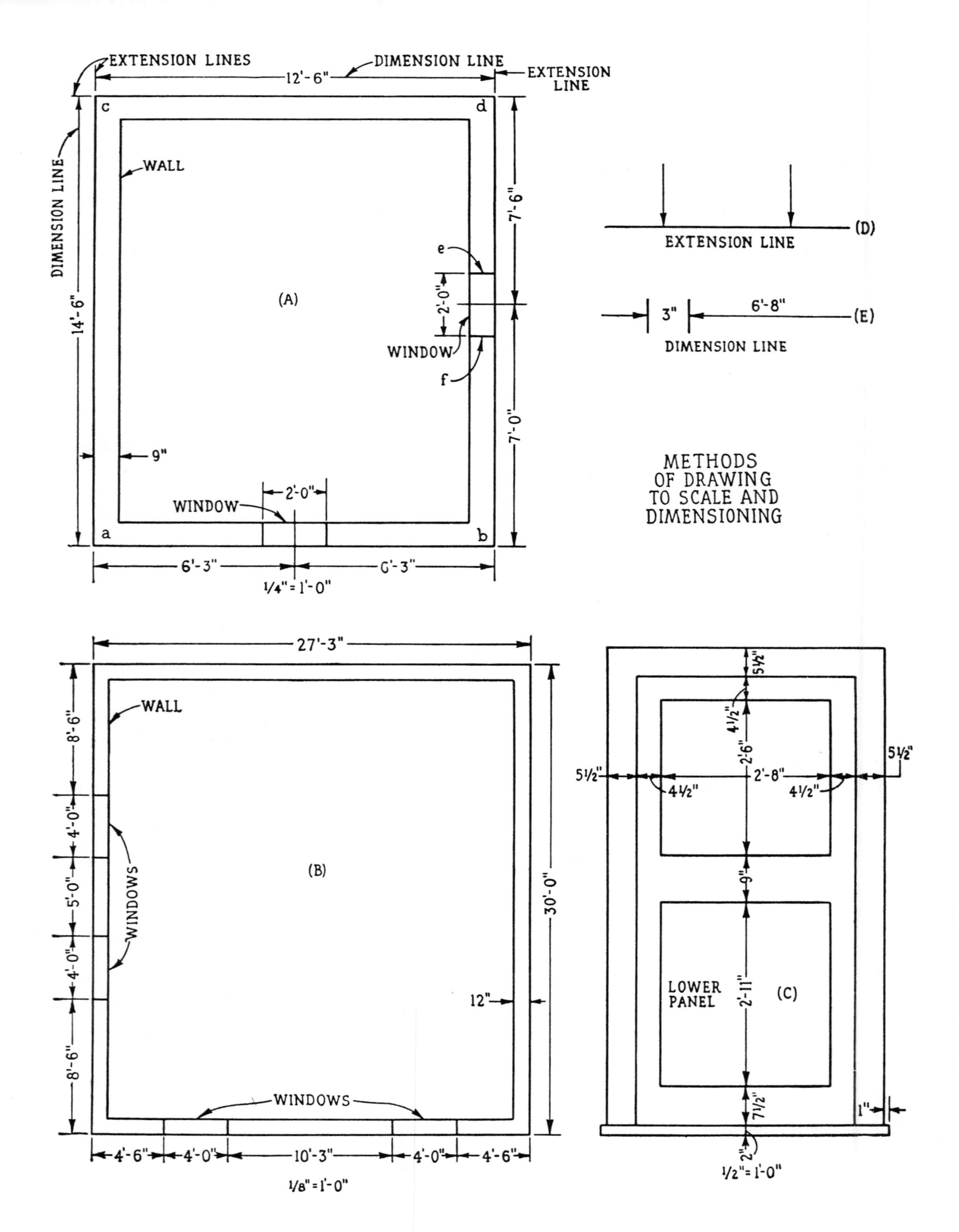
EXTENSION LINES
DIMENSION LINE
EXTENSION LINE
12'-6"
DIMENSION LINE
14'-6"
WALL
7'-6"
7'-0"
2'-0"
WINDOW
(A)
9"
2'-0"
6'-3"
6'-3"
1/4" = 1'-0"
EXTENSION LINE
(D)
3"
6'-8"
(E)
DIMENSION LINE
METHODS OF DRAWING TO SCALE AND DIMENSIONING
27'-3"
WALL
8'-6"
4'-0"
5'-0"
4'-0"
8'-6"
WINDOWS
(B)
30'-0"
12"
WINDOWS
4'-6"
4'-0"
10'-3"
4'-0"
4'-6"
1/8" = 1'-0"
5½"
4½"
2'6"
2'-8"
9"
LOWER PANEL
2'-11"
(C)
7½"
1"
2"
1/2" = 1'-0"

Figure 3

thickness must be shown. The window in wall *ab* is midway between *a* and *b*. Note that the two 6′ 3″ dimensions add up to exactly 12′ 6″. The window in wall *bd* is 7′ 0″ from *b* and 7′ 6″ from *d*. Note, also, that the sum of these two dimensions is exactly 14′ 6″. In other words, the total of all the dimensions used to locate or show sizes of items within the enclosure must equal the overall dimensions. No extension lines are needed to indicate the wall thickness. However, the arrows indicate the exact limits of the thickness dimension. The window widths, on the other hand, require extension lines, and the distance between these extension lines is exactly equal to the width of the windows. The symbols for extension and dimension lines are shown at *D* and *E* in Figure 3. In later chapters we shall see how extension and dimension lines are employed on drawings.

The enclosure shown in part *B* of Figure 3 was drawn using the ⅛″ = 1′ 0″ scale. Readers are urged to check all the dimensions using a three-sided scale or folding rule. Such practice will be of great aid in learning to visualize how drawings are made to scale and how to read them.

Note that in part *B* a different method of locating the windows was used. Designers have several slightly different procedures for the use of dimension lines, but the general principle is always the same. Both window and door openings in walls may be located by extension lines to their centers or sides.

Part *C* of Figure 3 shows a vertical drawing, or elevation view, which might be used as a picturelike representation of a door. This method of showing dimensions is typical for doors and other items where special millwork is concerned. That is, if a door of other than stock size and design is required, the architect makes drawings of this type and shows exactly what he has in mind. Note that the size of each part of the door can be determined from the vertical and horizontal rows of dimensions.

Sometimes designers use somewhat different types of dimension lines. As an example, note the difference between those in part *B* of Figure 2 and those in part *E* of Figure 3. Other variations, such as different types of arrows, are also used. However, the differences are never so great as to cause any visualization troubles.

MISSING DIMENSIONS

In many cases, and for one uncertain reason or another, designers do not show all the dimensions which estimators and builders actually need. It is seldom that important dimensions are omitted but most drawings are guilty of several dimension omissions. To determine what these dimensions are, estimators and builders are forced to scale the drawings.

Scaling In a previous section of this chapter we learned how to determine distances or spaces according to the scales specified. That same process, as explained in connection with part *B* of Figure 2, is employed to find missing dimensions on drawings.

EXAMPLE 1: Note the window whose sides are marked *f* and *e* in the *A* part of Figure 3. Suppose that the 2′ 0″ width dimension had not been shown. The exact width can be found by the use of a three-sided scale or a folding rule. If a three-sided scale is used, place the ¼″ = 1′ 0″ edge next to the window representation. Move the scale until the zero on it is directly over the side of the window marked *f*. The width of 2′ 0″ can be read from the scale. If a folding rule is used, simply place a portion of it along the window and count the quarter inches between *f* and *e*, just below the edge of the rule.

EXAMPLE 2: Imagine that other dimensions are missing in the *B* part of Figure 3. Go through the process of finding what the dimensions are, as explained in Example 1. This will be good practice.

Danger of Scaling Under various atmospheric conditions, print paper is likely to shrink. When that happens, the distances between lines are shortened and the scaling process becomes inaccurate. If a missing dimension is important, the best policy is to check with the designer or his original drawings.

REDUCED WORKING DRAWINGS IN TEXTBOOKS

When architectural drawings are reproduced in textbooks, it is often necessary to reduce them in size so that they can be shown within page-size limitations. Most such drawings, even when drawn to small scales like ¼″ = 1′ 0″, are much larger than the size of book pages. Consequently, the drawings must be reduced in size. The reducing is done by a photographic process.

When a scaled drawing has been reduced in size, we cannot scale it as previously explained. However, if a portion of a ¼″ = 1′ 0″ scale is reduced, along with the drawing, scaling can be done by the use of a reduced scale.

Note part *E* of Figure 2. Textbook drawings often include such a reduced scale. To find a missing dimension, set a pair of dividers to the unknown distance. Then, without changing the set of the dividers, apply their points to the reduced scale. The distance between the points will be the dimension required. If dividers are not available, a piece of paper having a straight edge can be used. Put the edge of the paper along the unknown distance and mark the limits of that distance on the paper with a pencil. Then apply the marked edge to the reduced scale.

QUESTIONS AND ANSWERS

Follow the suggested directions given for the questions and answers in Chapter 1.

Question 1 Suppose that an ordinary ruler is used to draw a line 3½ inches long. What distance would that line represent according to the quarter-inch scale?
Answer 1 On the quarter-inch scale, each of the quarter inches represents 1 foot. In 3½ inches, there are 14 quarter inches. Therefore, the 3½-inch line constitutes 14 feet on the quarter-inch scale.

Question 2 Suppose we use an ordinary ruler to draw a line 2⅝ inches long. What distance would that line represent on the ⅛″ = 1′ 0″ scale?
Answer 2 On this scale, each eighth inch represents 1 foot. In 2⅝ inches there are 21 eighth inches. Therefore, the line would constitute 21 feet on the given scale.

Question 3 Suppose that we again use a common ruler and draw a line 3¼ inches long. What distance would that represent on the half-inch scale?
Answer 3 On the half-inch scale, each half inch represents 1 foot. Each quarter inch represents 6 inches. Therefore, the line would be equal to 6½ feet.

Question 4 What scale would probably be used when drawing a large plot plan? Why?
Answer 4 Either the eighth- or sixteenth-inch scale would most likely be used, in order to confine the drawing to a sheet of reasonable size.

Question 5 How can we tell what scale was used for a drawing?
Answer 5 By looking for the scale notation which is generally near the title of each drawing.

Question 6 Are dimensions given according to the actual size of a structure or according to the lengths of lines on scaled drawings?
Answer 6 All dimensions are always given in terms of the actual size of a structure.

CHAPTER THREE

Elevation Views

Previous chapters pointed out the fact that construction plans are prepared to small scale and that, as a result, all structural parts must be indicated by symbols. In like manner, and because not many of them can be drawn in picturelike form, materials must also be indicated by symbols. In addition to the symbols, many building terms and abbreviations must be employed to supplement the symbols.

A previous chapter also pointed out that there are several types of working drawings and that each type has its own family of symbols. In other words, a symbol used for elevation views cannot be used for plan or other views. With that fact in mind, this book presents a separate chapter for each type of drawing.

The purpose of this chapter is first, to show how to visualize elevation views. Following that, instruction, illustrations and explanations of all symbols, terms, and abbreviations are given.

HOW TO VISUALIZE ELEVATION VIEWS

As an introductory example on the visualization of elevation views, let us imagine that we can stand so as to face squarely and look at, one after the other, the four sides of an existing house. As we look at each side, we see what is called an *elevation view*. In other words, architects refer to the sides of a structure as elevations.

Next, let us imagine that we can take separate pictures of each of the four sides of the same structure. We would have another form of elevation views.

Finally, suppose that it becomes necessary for us to describe the elevation views of that same structure to one or more other persons. Without the pictures, we would encounter great difficulty trying to create good, and exactly the same, mental pictures for the other people. With the pictures, our description could be accomplished easily and in such a manner that the other people could see all they would have to know.

With the foregoing, we are able to start our visualization of elevation views and to understand that such drawings show the exterior sides of a structure as it is to appear after all structural work has been completed.

Most builders are likely to study plan views before they concern themselves with the exterior appearance of a structure. However, in our study of print reading, we are deviating from that procedure because elevations are more familiar and because we must learn to visualize such views before we can consider plan views.

The visualization of plan views includes more details than beginners are likely to suspect. For that reason, we shall review several more or less fundamental concepts in order to establish a firm foundation for visualization.

Examples The *A* part of Figure 4 shows an object which has four sides. If we imagine that we are facing squarely the *abcd* side of the object and looking at it from the point *x*, we see the elevation view pointed to by the dashed-line arrow. Note that the elevation view includes the same *abcd* area as shown on the object. In like manner, if we imagine that we are facing the *cdfe* side of the object and looking at it from the point *y*, we see another elevation view as indicated by the dashed-line arrow. Note that the elevation view includes the same *cdfe* area as shown on the object. If we imagine that we can face the two sides of the object which are not shown in the illustration, we see the other two elevations of it.

The *B* part of Figure 4 shows the same object with a rooflike addition. If we imagine that, as before, we look at the object from the points *x* and *y*, we see the *x* and *y* elevations. Note that elevation at *x* includes the same *abcdg* area as shown on the object and that the elevation at *y* includes the same *cdghfe* area as shown on the object.

The *C* part of Figure 4 has some window- and doorlike indications added, to make the object look something like a house. The *x* and *y* elevations are visualized in the same manner as previously explained.

The *D* part of Figure 4 introduces a more irregularly shaped object which is something like a house having an ell on one side. Shapes of this kind are common in modern houses. We visualize the elevations for this object as previously explained. However, the process is a little more complicated because of the irregular shape.

Suppose we imagine that we are looking at the object from the point *x*. From that position, we can see the sides marked 1 and 2. We can also see the corners marked *ac, bd,* and *kg*. The corner *ef* is hidden by *bd* and the corner *jh* is hidden by *kg*. Thus, the elevation at *x* shows only the sides marked 1 and 2 and the corners marked *ac, bd,* and *kg*.

If we imagine that we are looking at the object from the point *y*, we see the sides marked 3 and 4 and the corners marked *bd, kg,* and *jh*. The corner *ef* is hidden by *kg* and the back corner is hidden by *jh*. Thus, the elevation at *y* shows only the sides 3 and 4 and the corners *bd, kg,* and *jh*. It should be noted that while some sides, such as the one marked 3, are not visible in one elevation, such as the one at *x*, they are visible in another, such as *y*. From this observation, we can understand that all four elevations of a structure must be studied before we can visualize true shape and appearance.

The *E* part of Figure 4 shows an object which is even more in keeping with the shapes of modern buildings. Suppose we imagine that we are looking at the object from the point *x*. We can see the side marked 1 and part of the side marked 2. We can also see the corners marked *ac, bd,* and *mk*. The corner *fg* is hidden by *bd* and *on* is hidden by *mk*. The corners marked *e* and *hj* are not visible. Thus the elevation at *x* shows the sides 1 and 2 and the corners *ac, bd,* and *mk*. The dashed line in the elevation view, under the letter *e*, represents

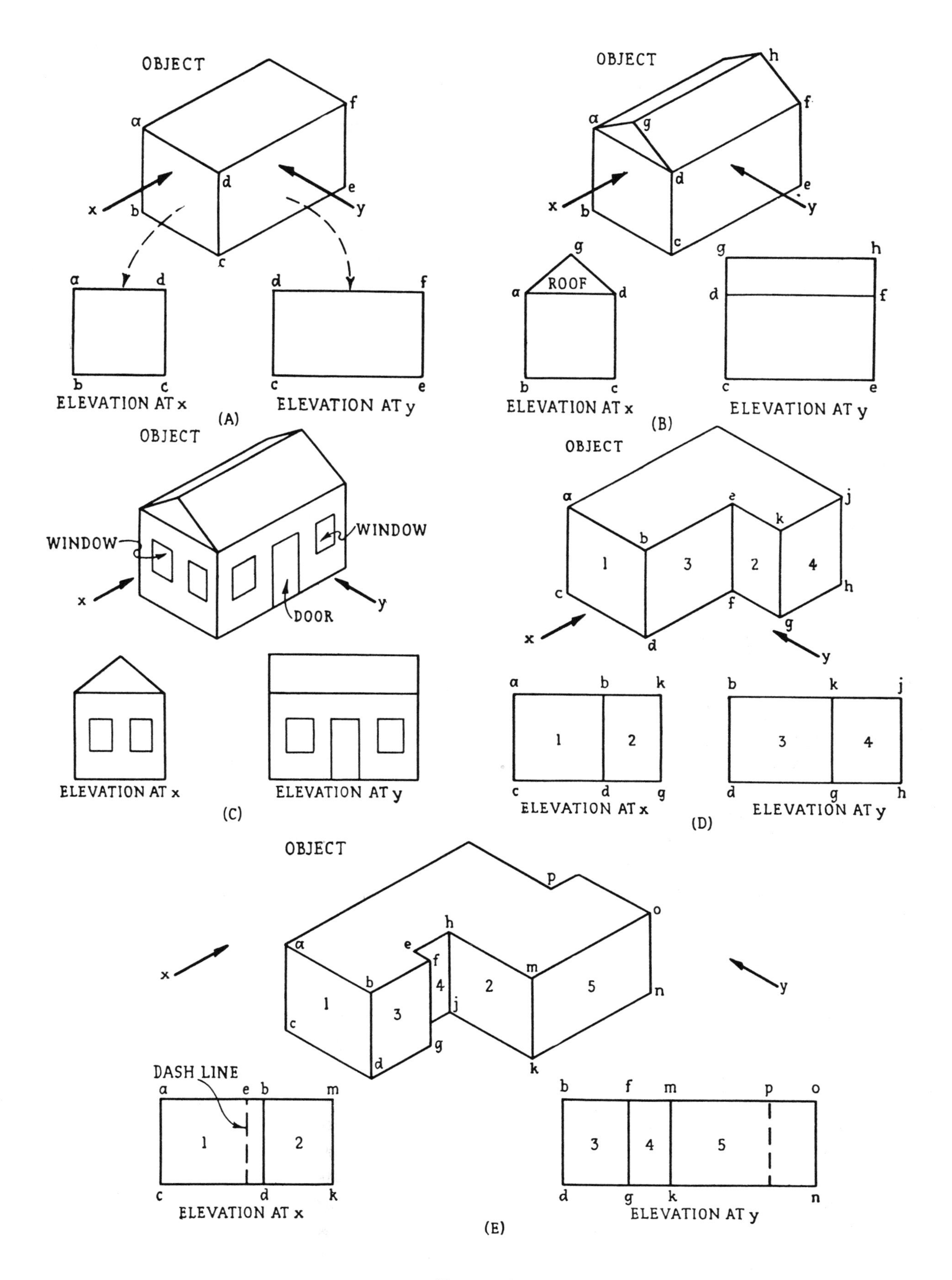
OBJECT
a
f
d
e
x
y
b
c
a
d
b
c
ELEVATION AT x
d
f
c
e
ELEVATION AT y
(A)
OBJECT
h
f
a
g
d
e
x
y
b
c
g
ROOF
a
d
b
c
ELEVATION AT x
g
h
d
f
c
e
ELEVATION AT y
(B)
OBJECT
WINDOW
WINDOW
x
DOOR
y
ELEVATION AT x
ELEVATION AT y
(C)
OBJECT
a
e
j
k
b
1
3
2
4
c
h
f
x
g
d
y
a
b
k
1
2
c
d
g
ELEVATION AT x
b
k
j
3
4
d
g
h
ELEVATION AT y
(D)
OBJECT
p
o
x
h
y
a
e
f
m
4
2
5
b
n
1
j
3
c
g
d
k
DASH LINE
a
e
b
m
1
2
c
d
k
ELEVATION AT x
b
f
m
p
o
3
4
5
d
g
k
n
ELEVATION AT y
(E)

Figure 4

the edge of the side marked 4 in the object. Such dashed lines are not used in ordinary elevation views, and we use this one only to aid in visualization.

If we imagine that we are looking at the object from the point *y*, we can see the sides marked 3, 4, and 5 and the corners *bd, fg, mk,* and *on*. The corner at *e* is hidden by *fg* and the corner at *hj* is hidden by *mk*. The dashed line in the elevation at *y* represents the corner *p* in the object. This object, while again indicating that four elevation views are necessary in order to visualize the shape and appearance of a structure, also makes clear the fact that structures having complicated or irregular shapes would be difficult to visualize without the aid of plan views. The relationship between plan and elevation views is explained in Chapter 4.

Elevation Views of Houses The *A* part of Figure 5 shows a perspective view of a single-story house. We are not yet familiar with symbols; therefore, the roof, sides, chimney, garage door, windows, and regular entrance door are shown and noted in the illustration.

The drawings for such a house would include four elevation views as shown in the *B, C, D,* and *E* parts of Figure 5.

The elevation view shown in the *B* part of the illustration is what we would see if we could look at such a house from the west side of it. The roof areas marked 6 and 8 are the same areas shown in the perspective view. The chimney area, marked 7, is the same as the side marked 7 in the perspective. In like manner, the roof points marked *e* and *f* are the same points shown in the perspective. The doors and windows marked 1, 2, 4, and 5 are the same as the like numbers shown in the perspective. Note how the perspective points *c* and *a* appear in the elevation view.

The elevation view shown in part *C* of the illustration is what we would see if we looked at the south side of the house. Note how the chimney extends above the top of the roof, that the roof area marked 9 is visible, and that the corner marked *gb*, in the perspective, appears at the left-hand side of the elevation view.

The elevation views shown in the *D* and *E* parts of the illustration can be visualized by the aid of the numbers which appear in both the perspective and elevation views.

Names of Elevation Views In order to name elevation views so that they can be referred to without confusion, they are often called north, south, east, and west elevations, as shown in Figure 5. In other cases, designers sometimes refer to elevations as front, rear, left-side, and right-side elevations.

SYMBOLS

What we shall illustrate and explain as *symbols* includes two subdivisions of the general term which are sometimes referred to as *symbols* and *conventions*. Actually, there is little difference in meaning between the two subdivisions as they are applied to construction plans. Precisely speaking, *symbols* are a means of indicating various kinds of building materials, and *conventions* constitute a method of presenting drawing details and parts of them, such as dimensions, in a standard way. In order to simplify the process of learning to read and understand construction plans, we shall use *symbols* throughout this book.

In this chapter, only the symbols used for elevation views are illustrated and explained. The symbols employed for other types of working drawings are explained in the chapters devoted to such other views.

SYMBOLS USED IN ELEVATION VIEWS

In order to create elevation views which are picturelike in appearance, designers draw materials, windows, doors, etc., to look as much like such items as possible. On the small scales necessary for the drawings, none of the items, especially materials, can be drawn exactly as

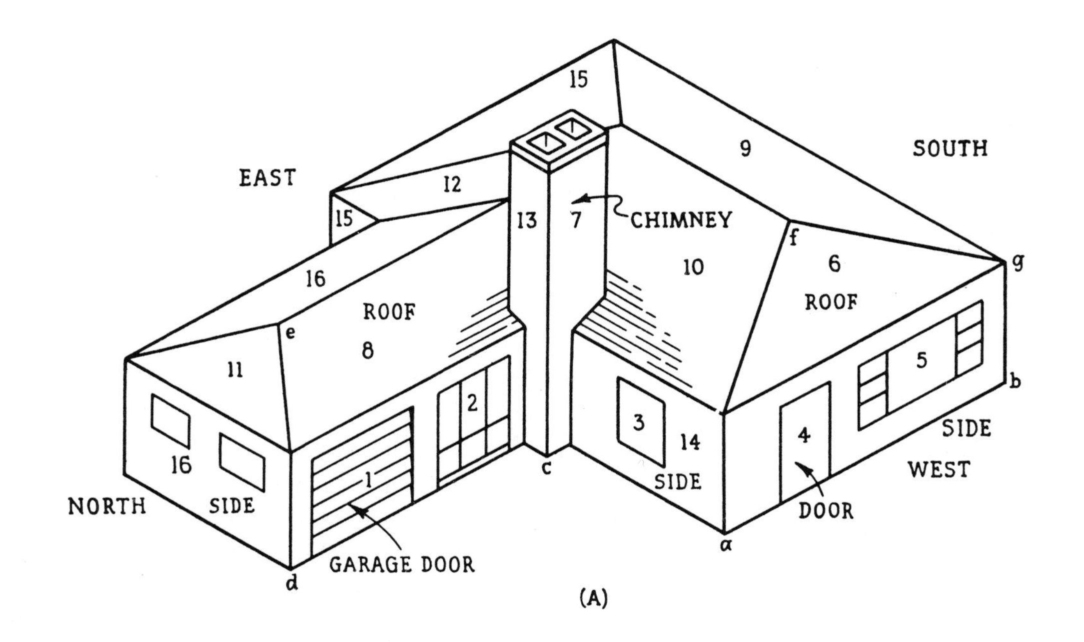

(A)

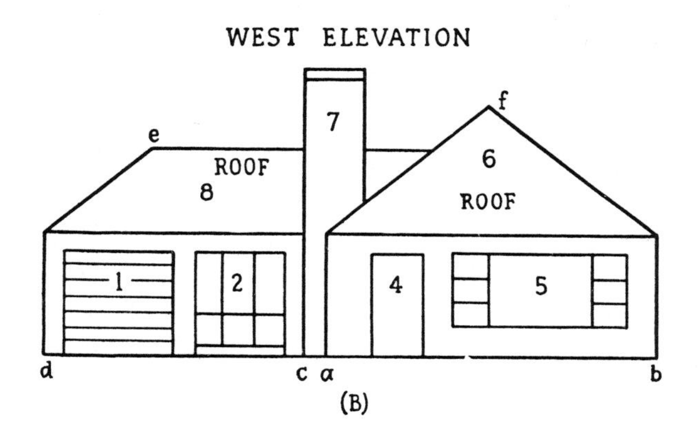

(B)

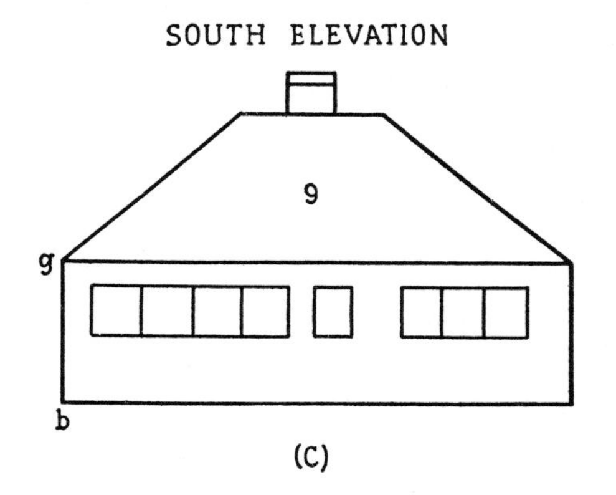

(C)

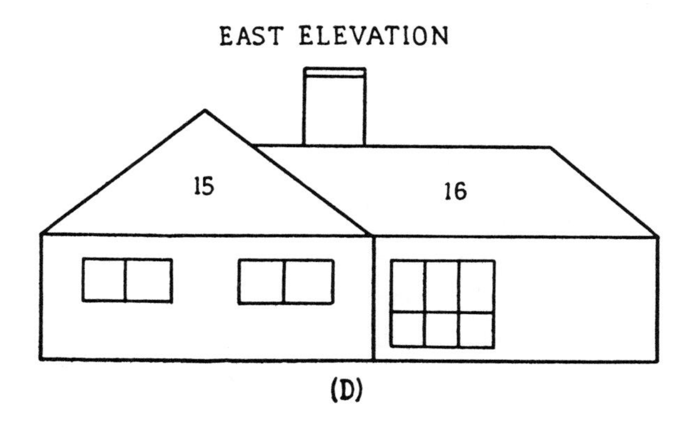

(D)

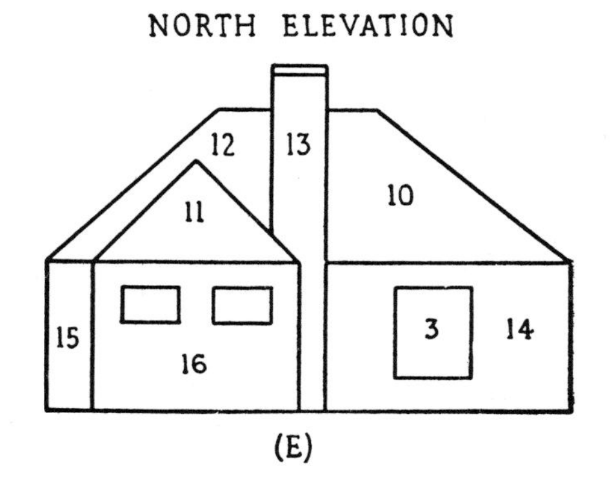

(E)

Figure 5

they appear. Therefore, representations and reasonably similar delineations, known as *symbols,* are necessary.

Material Symbols Some materials, such as brick or concrete blocks, can be drawn to scale and to look exactly like these materials. However, at the small scales employed, such drawings would require a great deal of expensive drafting-board time. Other materials, such as concrete, plaster, and stucco, are impossible to draw or even take a picture of. Thus, it is more economical, and much easier, to represent all kinds of materials by symbols.

Figure 6 shows a group of typical symbols which are used to represent commonly encountered building materials. We shall discuss a few of them in terms of what they represent and how they are drawn.

NOTE: We shall use the abbreviation *specs* to mean written specifications.

Brick The *A* part of Figure 6 shows the symbol which is almost always used to indicate brick. The symbol is drawn as a group of horizontal and parallel lines over all, or just part of, the areas to be constructed of brick. As a means of saving drafting-board time, designers frequently draw this symbol only in small patches rather than cover all brick areas with it. It does not indicate the kind or color of brick or the type of mortar to be used. That sort of information is always given in the specs. The symbol is drawn fairly accurately to scale. At the ¼″ = 1′ 0″ scale, for example, the distance between the horizontal lines, as indicated by the dimension *a* in the illustration, would be about 1⁄16 inch.

Concrete The *B* part of the illustration shows the symbol almost always used to indicate ordinary concrete. It is merely a group of dots. Instructions concerning mix are seldom shown on or near the symbol but are described in the specs. No scale is involved.

Concrete Block The *C* part of the illustration shows a commonly used symbol for this material. It is drawn just about as the blocks appear in a wall. However, the symbol does not indicate the kind of concrete, the kind of mortar, or the joint thickness. All such information is given in the specs. Concrete blocks are generally 7⅝ inches high and 15⅝ inches long. Thus, at the ¼″ = 1′ 0″ scale, each block in the symbol is drawn about 3⁄16 inch high and ⅜ inch long.

Board and Batten The *D* part of the illustration shows the symbol for this type of vertical wood siding. In the illustration, the *a* indicates a board width and the *b* indicates a batten width. Any one of several kinds of wood, as directed in the specs, can be used. The symbol is drawn to scale so far as the widths of boards and battens are concerned.

The *F*, *H*, and *L* parts of the illustration all employ the same dotted symbols as used to indicate concrete. This is because the materials being represented, like concrete, are difficult to picture. In all cases, the specs provide any information which the symbols cannot show.

The *M* part of the illustration shows how horizontal wood siding is indicated. The horizontal lines represent individual pieces of siding. Any one of several kinds and sizes of wood, as directed in the specs, can be used. The horizontal lines are generally drawn to scale at a distance apart which will show how much each piece of siding is exposed to the weather.

The *O* and *P* parts of the illustration show alternate ways of indicating wood shingles. The kind of wood of which the shingles are made is generally set forth in the specs. The symbols are drawn fairly well to scale, so far as the exposure to the weather is concerned.

The *Q* and *R* parts of the illustration show how various kinds of roofing materials are sometimes specified on elevation views.

All other material symbols shown in the illustration are drawn, specified, or scaled as explained for the foregoing symbols.

Uncommon Symbols Sometimes, uncommon materials are to be used for various structural parts of a structure. In such cases, designers generally include a *symbol key* in one or more

ELEVATION VIEW MATERIAL SYMBOLS

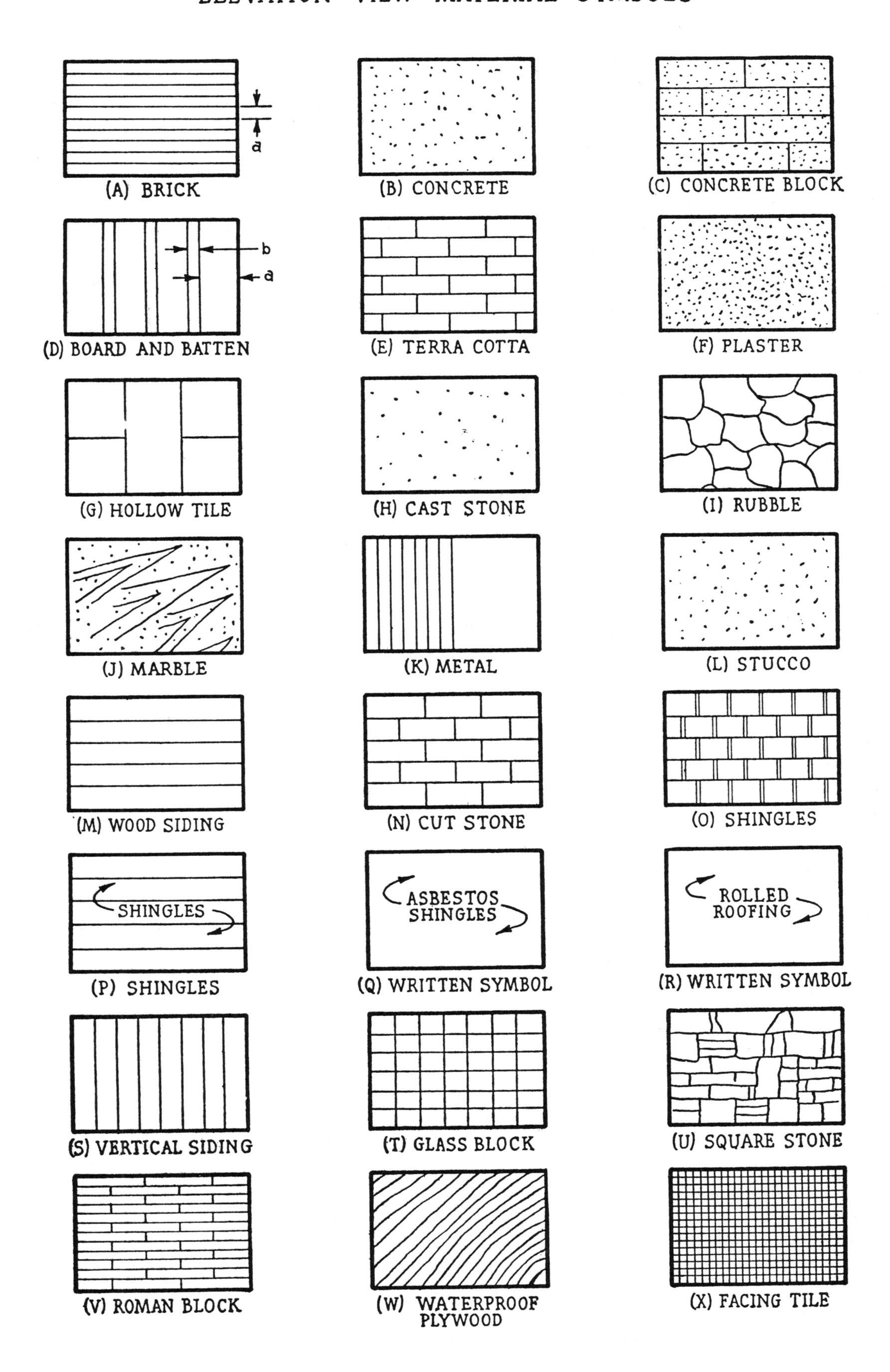

Figure 6

of the various drawings. Such a key gives uncommon symbols together with the names of the materials shown.

Detail Symbols As previously explained, such details as windows and doors are represented by delineations which are similar to the actual objects. The delineations serve the purpose and save a great deal of drafting-board time.

Figure 7 shows a group of symbols for windows, doors, and a typical louver.

The *A* part of Figure 7 shows five window symbols which are lettered *a*, *b*, *c*, *d*, and *e*.

At *a*, the double-hung window has twelve 7- by 9-inch panes of glass and shutters. It is shown in a brick wall.

At *b*, the same kind and size of window, with shutters, is shown in a wood siding, or frame, wall.

At *c*, a like window, without shutters, is shown in a brick wall.

At *d*, a casement-type window having sixteen 5- by 7-inch panes of glass is shown in a stucco wall.

At *e*, another double-hung window having twelve 7- by 9-inch panes of glass is shown in a wood siding, or frame, wall.

All other needed information about the windows can be found in the specs.

The *B* part of the illustration shows the symbol for a picture window which has double hung windows on both sides of it. The sizes of glass are sometimes given in the symbol and at other times in the specs. All information relative to kind or grade of glass, kind of wood, etc., is also to be found in specs.

The *C* part of the illustration shows a typical symbol for a picture window which is flanked by awning-type windows. Architects sometimes include circled letters with the symbols which refer either to the specs or to a window schedule which appears on one of the drawings. The schedule provides whatever information is not given in the specs.

All window symbols are drawn closely to scale.

The *D* part of the illustration gives the symbol for a typical kind of louver. Note that the louver is shown in a shingle wall.

The *E* part of the illustration shows how various kinds of doors are indicated in elevation views. Such symbols are sometimes shown with frames around them in much the same manner that the windows are shown. The size and grade of glass are noted on the symbols or given in the specs. Almost without exception, the kind of wood, type of construction, and thicknesses are given in the specs.

Other Detail Symbols In addition to the symbols explained in connection with Figure 7, there are many more, frequently used as parts of elevation views. In all cases, the symbols are drawn to look like the details they represent. Among designers, there is a rather large range of methods employed to give all the information needed in connection with such details. However, in one way or another, all information is available and can be found by searching through the drawings and specs.

Variations in Symbols In various sections of the country symbols are apt to vary somewhat in appearance. For example, in California designers may use a somewhat different symbol, for a given detail, from that used in New York. However, the possible differences are not great enough to cause serious confusion once we have learned to read and visualize drawings. In the event that we do encounter a foreign symbol, we can inquire as to its meaning.

Why Should a Mechanic Be Able to Read and Understand All Drawing Symbols? In general, where work involves two or more trades, the mechanic should know what parts are concerned and what, if any, provisions must be made by him for tying in his trade with the work of others. Also, unless a mechanic can recognize all symbols, he cannot visualize a drawing properly.

TYPICAL ELEVATION VIEW DETAIL SYMBOLS

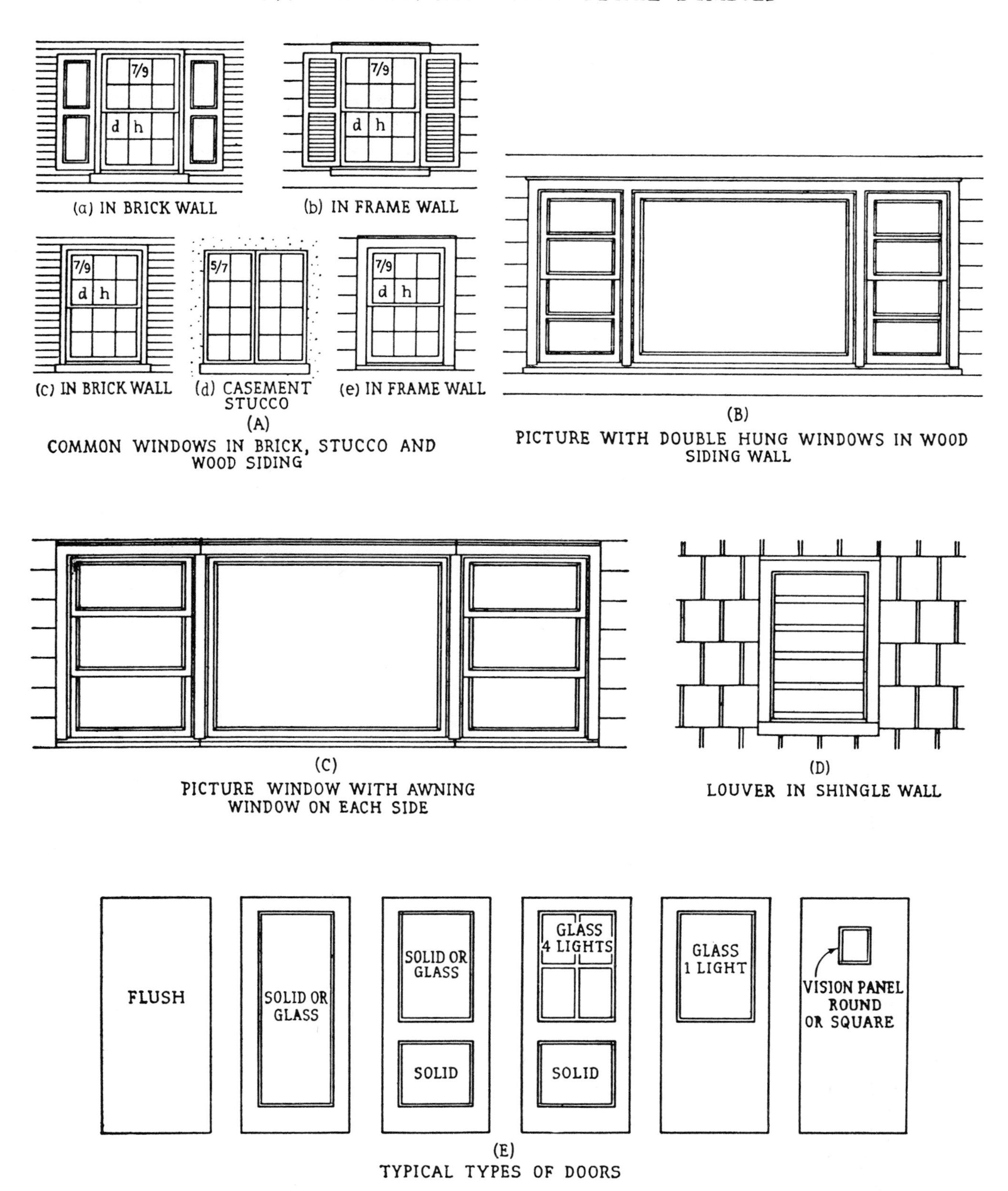

Figure 7

TERMS

APRON: The finish board placed immediately below a window stool (see *A* part of Figure 9).

ARCHITRAVE: The molded finish around an opening, also called *trim*.

ASHLAR: The outside cut-stone facing of a wall.

AWNING WINDOW: Separate windows or sections of window walls which open out, similar to an awning (see *B* part of Figure 9).

BATTEN: A strip of wood for fastening other boards together.

BOARD AND BATTEN: Vertical wood siding composed of wide boards with battens over the cracks between boards.

CASEMENT: A hinged window.

CEMENT PLASTER: Mortar made with various mixes of cement and sand used as fireproof plaster or as waterproofing on the exteriors of foundations.

CLAPBOARD: A siding which is long and thin and graduated in thickness.

CLERESTORY WINDOWS: Small windows used under roofs where one section of roof is higher than another section. The windows are in the walls between such roofs.

CONCRETE BLOCK VENT: A vent or louver made to the exact size of a concrete block and used in walls made of such block (see *E* part of Figure 9).

CORBEL: A bracket formed in a wall by building out successive courses of masonry.

CORNER BOARDS: Corner trim for wood siding or frame wall (see *B* part of Figure 8).

CORNICE: The part of a roof that projects beyond a wall.

CORNICE RETURN: Short portion of cornice carried around corner of house (see *B* part of Figure 8).

COURSE: A continuous row of brick or other masonry units.

CROWN MOLDING: A molding at the top of a cornice and just under the roof (see *H* part of Figure 9).

DORMER: A structure projecting from a sloping roof. It is a means of providing light and air for areas under roofs (see *C* part of Figure 8).

DRIP: A construction member which projects, as above windows, to throw off water (see *B* part of Figure 9).

DRIP MOLD: A molding designed to protect wall faces from water which would otherwise run down them.

FASCIA: The outside, flat member of a cornice (see *G* part of Figure 9).

FINISH STRIP: Part of a cornice which is used with a crown molding (see *H* part of Figure 9).

FINISHED CEILING: A term used on elevation views to indicate the ceiling height.

FINISHED FIRST FLOOR: A term used on elevation views to designate the position of the first floor level.

FIXED WOOD SASH: Windows, such as window walls or picture windows, which do not open.

FLASHING: The sheet metal work which prevents leakage over windows, doors, etc., around chimneys, and at intersecting parts (see *D* part of Figure 8).

FLAT ROOF: A roof which has just enough pitch to permit water to run off.

FRIEZE: A trim member used just below the cornice (see *H* part of Figure 9).

GABLE: The triangular portion of an end wall formed by the sloping roof.

GABLE ROOF: A roof which slopes up from two walls.

GAMBREL ROOF: A roof which has two different slopes.

GRADE: The level of the ground around a building (see *I* part of Figure 9).

GUTTER: A trough or depression for carrying water from roofs (see *G* part of Figure 8).

HIP ROOF: A roof which slopes up from all walls of a building.

HOOD: A small roof over a doorway (see *D* part of Figure 9).

JAMB: The inside vertical face of a window or window frame (see *A* part of Figure 8).

LALLY COLUMN: A cylindrically shaped steel member, sometimes filled with concrete, used as a support for girders or other beams.

LEADER: A metal pipe used to carry roof water from the gutter to the ground (see *G* part of Figure 8).

LINTEL: The horizontal structural member supporting a wall over window or door openings.

LOOKOUT: A short timber for supporting a projecting cornice.

LOUVER: An opening for ventilation covered by sloping slats to exclude rain (see *F* part of Figure 9).

MEETING RAIL: The horizontal center rails of a sash in double hung windows (see *A* part of Figure 8).

METAL CAPS: Metal flashings over windows and doors to make them waterproof.

MILLWORK: Wooden parts which are finished and partly assembled at a woodworking mill.

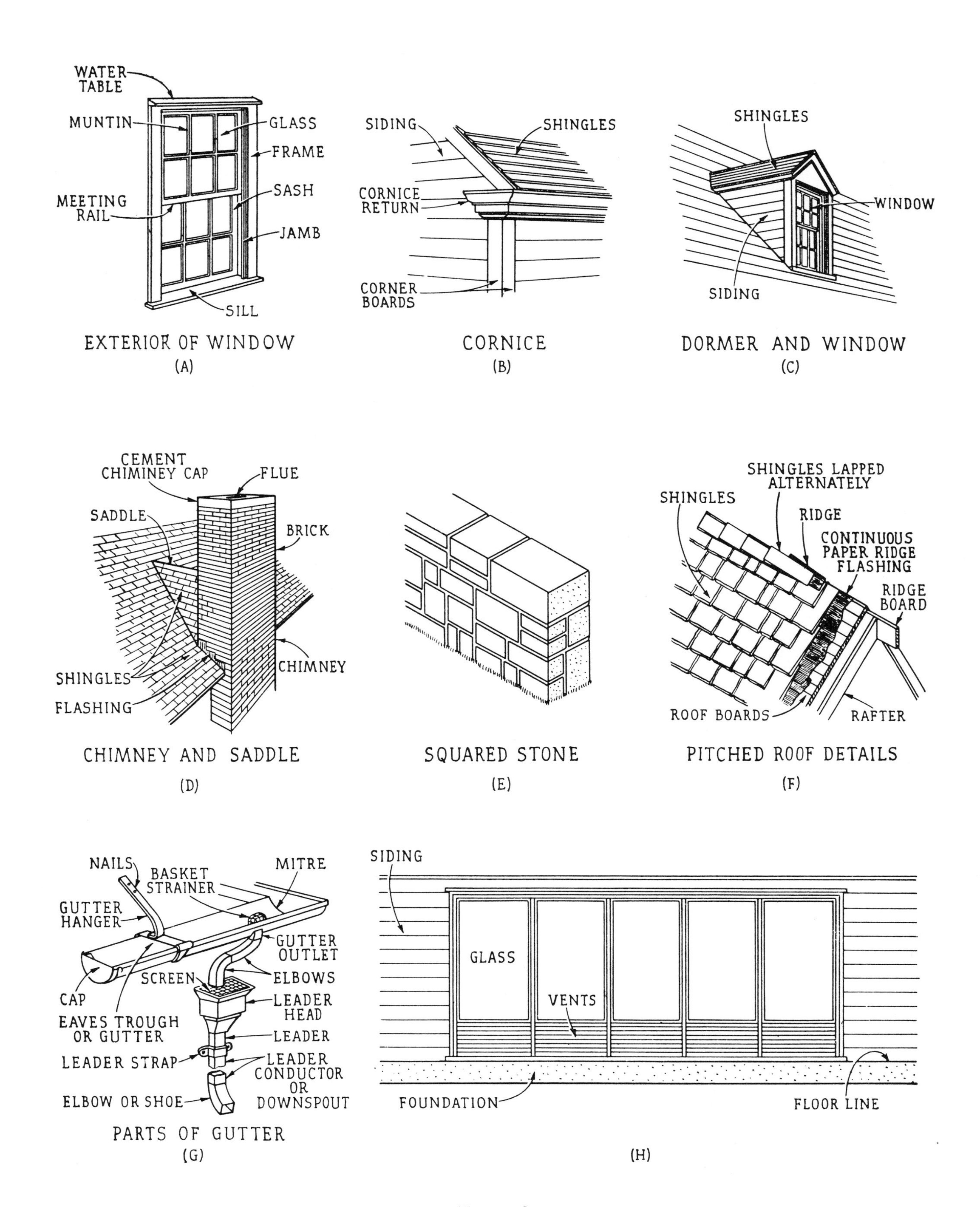
WATER TABLE
MUNTIN
GLASS
FRAME
MEETING RAIL
SASH
JAMB
SILL
EXTERIOR OF WINDOW
(A)
SIDING
SHINGLES
CORNICE RETURN
CORNER BOARDS
CORNICE
(B)
SHINGLES
WINDOW
SIDING
DORMER AND WINDOW
(C)
CEMENT CHIMINEY CAP
FLUE
SADDLE
BRICK
CHIMNEY
SHINGLES
FLASHING
CHIMNEY AND SADDLE
(D)
SQUARED STONE
(E)
SHINGLES LAPPED ALTERNATELY
SHINGLES
RIDGE
CONTINUOUS PAPER RIDGE FLASHING
RIDGE BOARD
ROOF BOARDS
RAFTER
PITCHED ROOF DETAILS
(F)
NAILS
BASKET STRAINER
MITRE
GUTTER HANGER
GUTTER OUTLET
ELBOWS
SCREEN
CAP
LEADER HEAD
EAVES TROUGH OR GUTTER
LEADER
LEADER STRAP
LEADER CONDUCTOR OR DOWNSPOUT
ELBOW OR SHOE
PARTS OF GUTTER
(G)
SIDING
GLASS
VENTS
FOUNDATION
FLOOR LINE
(H)

Figure 8

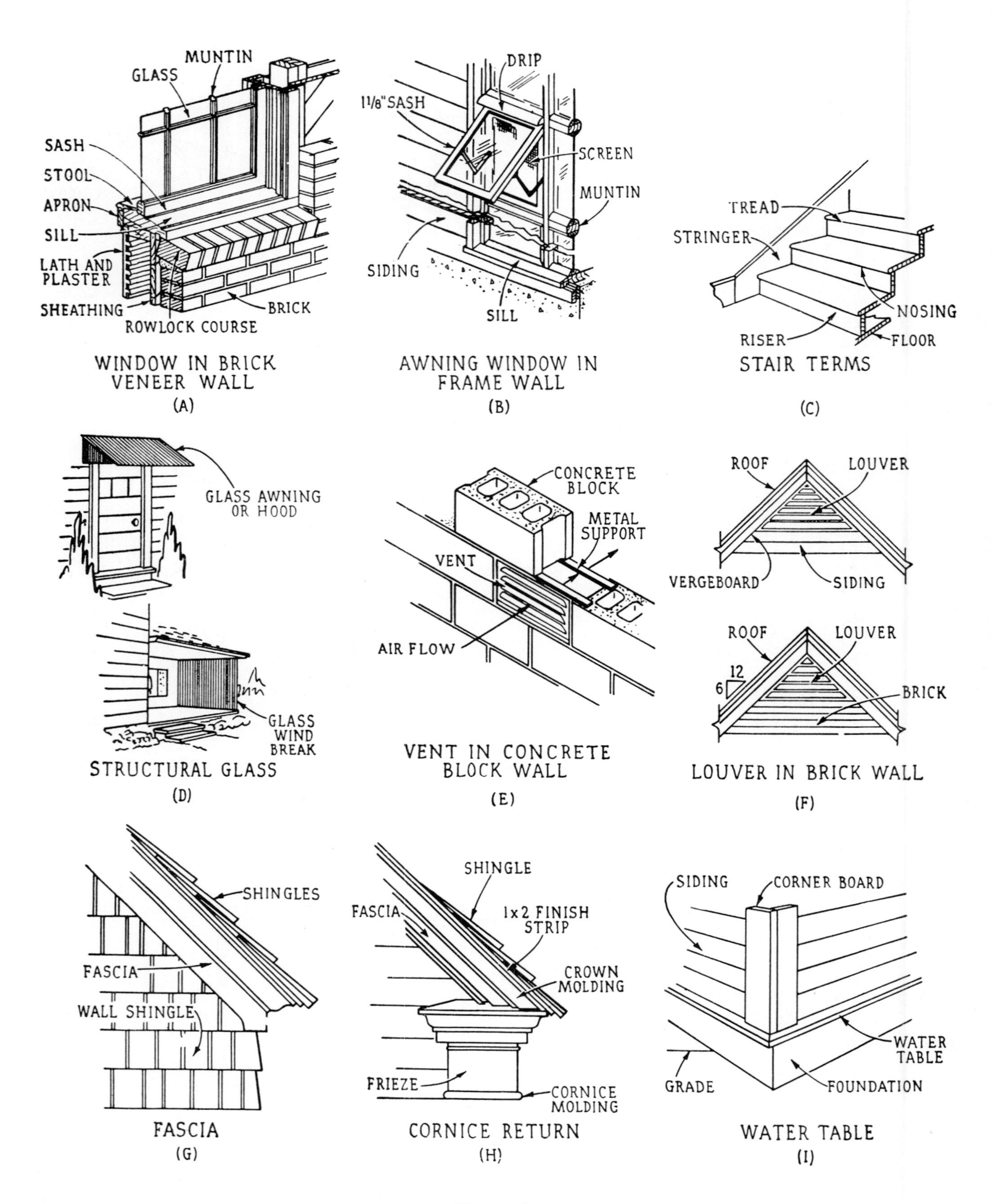
MUNTIN
GLASS
SASH
STOOL
APRON
SILL
LATH AND PLASTER
SHEATHING
ROWLOCK COURSE
BRICK
WINDOW IN BRICK VENEER WALL
(A)
DRIP
1 1/8" SASH
SCREEN
MUNTIN
SIDING
SILL
AWNING WINDOW IN FRAME WALL
(B)
TREAD
STRINGER
NOSING
RISER
FLOOR
STAIR TERMS
(C)
GLASS AWNING OR HOOD
GLASS WIND BREAK
STRUCTURAL GLASS
(D)
CONCRETE BLOCK
METAL SUPPORT
VENT
AIR FLOW
VENT IN CONCRETE BLOCK WALL
(E)
ROOF
LOUVER
VERGEBOARD
SIDING
ROOF
LOUVER
12
6
BRICK
LOUVER IN BRICK WALL
(F)
SHINGLES
FASCIA
WALL SHINGLE
FASCIA
(G)
SHINGLE
FASCIA
1x2 FINISH STRIP
CROWN MOLDING
FRIEZE
CORNICE MOLDING
CORNICE RETURN
(H)
SIDING
CORNER BOARD
WATER TABLE
GRADE
FOUNDATION
WATER TABLE
(I)

Figure 9

MITER: The beveled surface cut on the ends of a molding.

MULLION: The large vertical division of a window opening.

MUNTIN: A strip of wood which separates the panes of glass in a window frame (see *A* part of Figure 8).

NOSING: The overhanging edge of a stair tread (see *C* part of Figure 9).

PANEL: A piece of wood framed by other pieces.

PARTING STRIP: The strip in a double-hung window frame that keeps the upper and lower sash apart.

PITCH OF ROOF: A term applied to the amount of slope (see *F* part of Figure 9).

RAKE BOARD: See Verge Board.

RIDGE: The top edge of a roof where the two slopes meet (see *F* part of Figure 8).

RISER: The vertical portion of a step (see *C* part of Figure 9).

ROWLOCK COURSE: Course of brick set on edge under a window or other opening (see *A* part of Figure 9).

R. W. SIDING: Siding made of redwood.

SADDLE: Portion of roof built up between a chimney and the main part of the roof to throw water away from chimney (see *D* part of Figure 8).

SASH: The parts of a window in which panes of glass are set; generally movable, as in double-hung windows (see *A* part of Figure 8).

SHEATHING: The rough boarding on the outside of a wall or roof over which is laid the finish siding or shingles or brick (see *A* part of Figure 9).

SIDELIGHTS: Fixed windows, such as those on both sides of a door.

SILL: The bottom member under a window or door (see *A* part of Figures 8 and 9).

SOFFIT: The under surface of a cornice.

SOLDIER COURSE: Bricks laid on end with edges exposed.

TYPICAL ABBREVIATIONS

Asphalt	Asph.
Bedroom	B.R.
Bevel Siding	B.S.
Brick	Br.
Building	Bldg.
Cased Opening	C.O.
Casement	Csmt.
Cast Iron	C.I.
Ceiling	Clg.
Cement	Cem.
Center Line	L
Clapboards	Clapbds.
Clear	Clr.
Column	Col.
Concrete	Conc.
Conductor	Cond.
Conduit	Cond.
Copper	Cop.
Cornice	Corn.
Crock Tile	C.T.
Crystal	Crys.
Diagonal	Diag.
Diameter	Diam.
Dimension	Dim.
Dining Room	D.R.
Ditto	Do.
Divided	Div.
Door	Dr.
Double Hung	D.H.
Down	Dn.
Downspout	D.S.
Drain	Dr.
Drawing	Drg.
Drawn Glass	D.G.
Drip Cap	D.C.
Drop Cord	Dr. C.
Each	Ea.
Elevation	El.
Entrance	Ent.
Exterior	Ext.
Finish	Fin.
Finished Ceiling	Fin. Ceil.
Flashing	Flash.
Floor	Fl.
Foot or Feet	Ft.
Footing	Ftg.
Galvanized Iron	G.I.
Garage	Gar.
Glass	Gl.
Grade	Gr.
Gypsum Board	Gyp. Bd.
Height	Ht.
High Point	H.P.
Light	Lt.
Living Room	L.R.
Low Point	L.P.
Molding	Mldg.
Mullion	Mull.
Number	No. or #
Obscure	Obs.
On Center	O.C.
Outside	O.S.
Outside Casing	O.C.
Plaster	Pl.
Plate	Pl.
Plate Height	Pl. Ht.
Radius	R.
Room	Rm.
Redwood	R.W.
Screen	Scr.
Siding	Sdg.
Specifications	Specs.
Terra Cotta	T.C.
Threshold	Th.
Tongued and Grooved	T and G
Typical	Typ.
Veneer	Ven.
Vertical Tongued and Grooved	V.T. and G.
Wall Plug (Outlet)	W.P.
Wall Switch	S.
Waterproof	Wp.
Wide	W.
Wire Glass	W.G.
Wood	Wd.
Wood Casing	W.C.
Wrought Iron	W.I.
Yard	Yd.

STOOL: The base or support at the bottom of a window (see *A* part of Figure 9).

STRINGER: The supporting timber at the sides of a staircase (see *C* part of Figure 9).

STRUCTURAL GLASS: A special kind of glass which can be obtained in colors, and used indoors or outdoors (see *D* part of Figure 9).

T AND G SIDING: Siding, generally applied vertically, which has tongues and grooves in the edges of the boards, as with flooring.

TREAD: The horizontal board of a stair (see *C* part of Figure 9).

TRIM: The finish around an opening.

VALLEY: The intersection of two roof slopes.

VERGE BOARD: The board running down the slope of a roof from the top of a gable (see *F* part of Figure 9).

WATER TABLE: A projection of masonry or wood on the outside of the foundation to protect the foundation from rain (see *I* part of Figure 9).

WEATHER STRIP: Metal, wood, or some other material used to cover the joints around windows and doors to prevent drafts and dirt from entering buildings.

WINDOW WALL: Large expanse of fixed glass windows which form practically one wall of a room (see *H* part of Figure 8).

WINDOW WATER TABLE: The projecting and sloping member over a window to throw water away from the window (see *A* part of Figure 8).

TYPES OF HOUSES

The houses being planned and built today can be divided into many types, in so far as their external appearance is concerned. In fact, we should experience great difficulty if we tried to group them into even a reasonable number of general types. With such a fact in mind, we shall confine our consideration and discussion of houses to a few popular *kinds* without regard for exact types. Thus, the following illustrations and explanations concern split-level, one-story, and two-story houses.

Split-level Houses The *A* part of Figure 10 shows a perspective view of a typical split-level house. If we consider the lawn, at *d*, to be the grade level, we can easily visualize the fact that the points indicated by arrows *a*, *b*, and *c* are at different levels. The *B* part of Figure 10 shows another way of visualizing how the various levels are related. In houses of this kind, the living area is slightly above grade and generally includes a living room, a kitchen, and a dining space. This area, or level, is indicated by the arrow *b* in the *A* part of Figure 10 and by the living level in the *B* part of the illustration. The sleeping area generally includes bedrooms and a bath and is up one flight of stairs from the living area. This area, or level, is indicated by the arrow *a* in the *A* part of Figure 10 and by the sleeping level in the *B* part of that illustration. The utility area generally contains a garage, heater room, and laundry, and is down one flight of stairs from the living area. This area, or level, is indicated by the arrow *c* in the *A* part of Figure 10 and by the utility level in the *B* part of that same illustration.

The activity areas of split-level houses are well defined and located. The sleeping space is separate from the rest of the house; the living area is closely integrated with the outdoors; the utility area is convenient to the first-floor work area.

The three areas, or levels, are complete within themselves as sections, without regard to other areas, or levels, or to exterior appearance. To visualize the elevation views of such houses, it is only necessary to keep in mind the various levels, as shown in the two parts of Figure 10, and to think of each area, or level, as a separate part of the general design. In the following, we shall discuss only the elevation views. However, in Chapter 4 we shall see how the plan views are visualized, along with the elevation views.

Figure 11 shows typical front and right-side elevation views for the same house as illustrated by the perspective view in the *A* part of Figure 10. To visualize the elevation views, in connection with the perspective view, let us recall

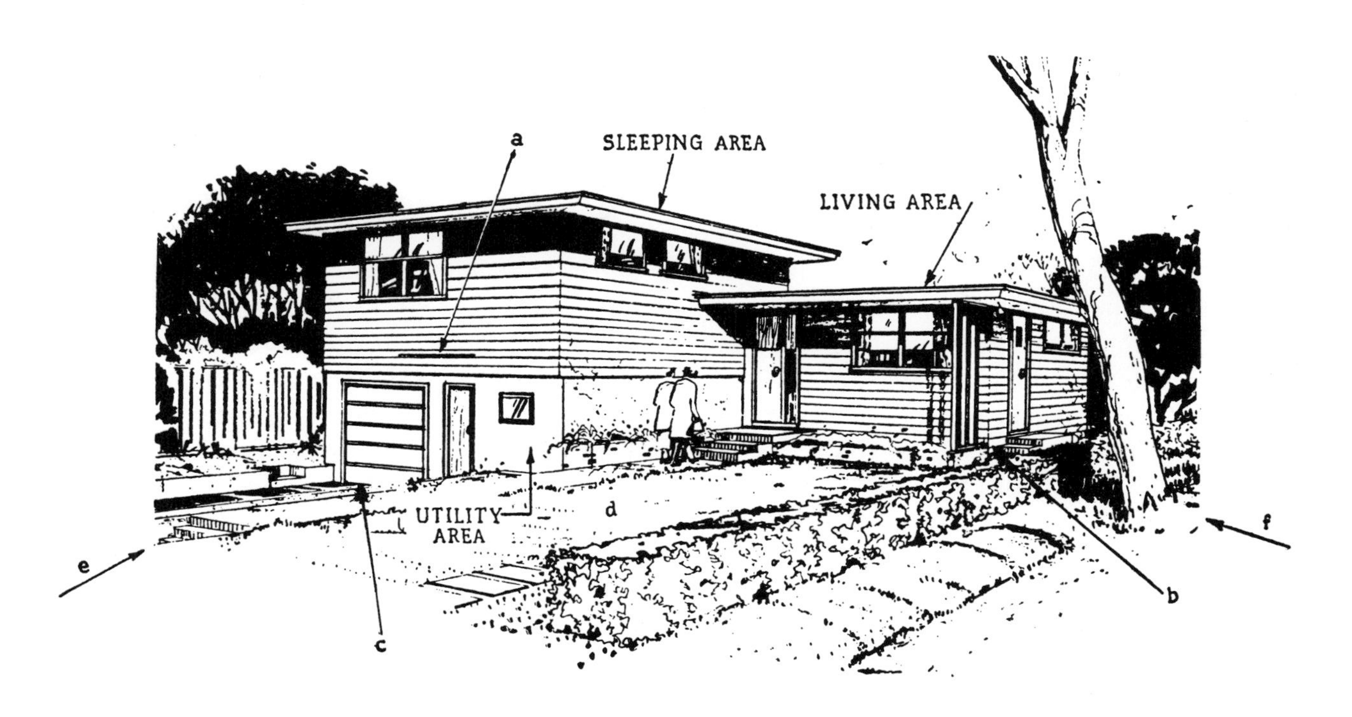

(A)
PERSPECTIVE VIEW

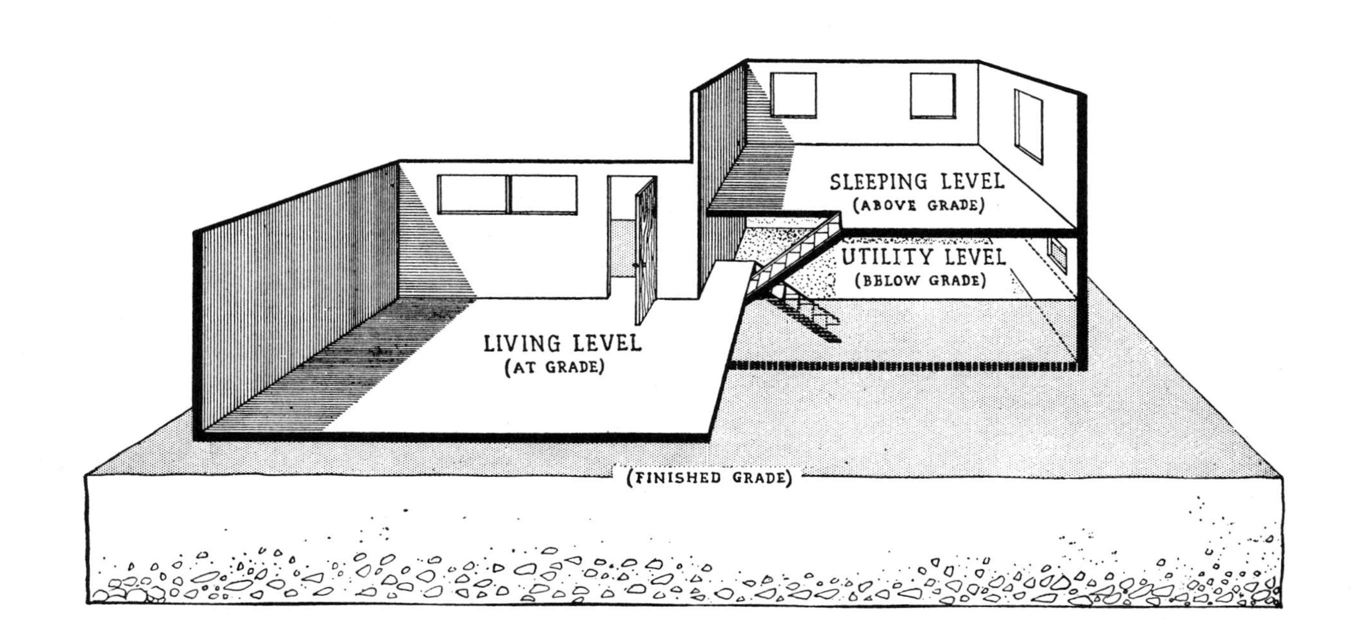

(B)
FLOOR LEVELS

Figure 10

the explanations given in regard to Figures 4 and 5 of this chapter. For example, let us imagine that we can look at the house, shown in the *A* part of Figure 10, from the direction of the arrow at *e*. What we see constitutes the front elevation as shown in Figure 11. Then, if we also imagine that we can look at the house from the direction of the arrow at *f*, what we see constitutes the right-side elevation. In other words, the elevation views in Figure 11 show the same two sides of the house as are illustrated by the perspective view in the *A* part of Figure 10.

As a further aid to visualizing the elevation views shown in Figure 11, note the number and positions of all doors and windows, on the front of the house, in both the perspective and front elevation views. For example, the *A* part of Figure 10 shows three ordinary doors and one garage door. In the front elevation view of Figure 11, two of those ordinary doors are marked *F* and *K* and the garage door is named. By using this same method of comparison with the elevation and perspective views, we can greatly improve our visualization ability.

Readers are urged to study the perspective and elevation views until they can visualize the house from the elevation views as easily as from the perspective view.

One-story Houses In one-story houses the living, sleeping, and utility areas are all on the same level. In other words, there is only one floor.

One-story houses may have a great variety of shapes and room arrangements, similar to the ones shown in Figures 4 and 5.

Figure 12 shows two elevation views for a typical one-story house. The general shape of the house is indicated by the small sketch which accompanies the elevation views. Note that this shape is somewhat similar to the shapes shown in the *D* part of Figure 4 and in the *A* part of Figure 5. Note, too, that the *A* part of Figure 5 is a one-story house.

Some one-story houses are known as ranch houses. Other special names are also used. However, once we learn to visualize the elevation views shown in Figure 12, we can visualize any and all types or kinds of houses having only one floor level.

Two-story Houses As the name implies, two-story houses have what are called first and second floors. In most cases, the first floor includes living room, kitchen, and dining space. The second floor includes bedrooms and baths. Such houses, unlike most one-story houses, ordinarily have basements where heater rooms and laundry facilities are located.

Figure 13 shows two elevation views for one type of two-story house. The house is almost square. Many two-story houses are rectangular. Some have a shape similar to the *D* part of Figure 4.

Two-story houses, such as represented by the elevation views in Figure 13, are often designed in keeping with what is known as traditional styling. Such styling, or types of houses, originated a long time ago and is divided into such classifications as Cape Cod, Old English, Early English, Georgian, Garrison, etc. However, once we have learned to visualize the typical elevation views in Figure 13, we can just as easily visualize elevation views for any type or kind of two-story house.

One-and-a-half-story Houses Sometimes houses are designed so that they have roofs which slant from the ceilings of their first floors to a peak or ridge. Thus, the spaces under the roofs are a great deal less in area than the first-floor spaces. In many cases, the space under the roof is used for two or more bedrooms. In such cases, dormers, larger than those shown in Figure 13, are required as a means of giving some additional height to ceilings. The dormers also make two or more windows possible. Such houses are called *one-and-a-half-story houses* because the second floor has much less area than the first floor.

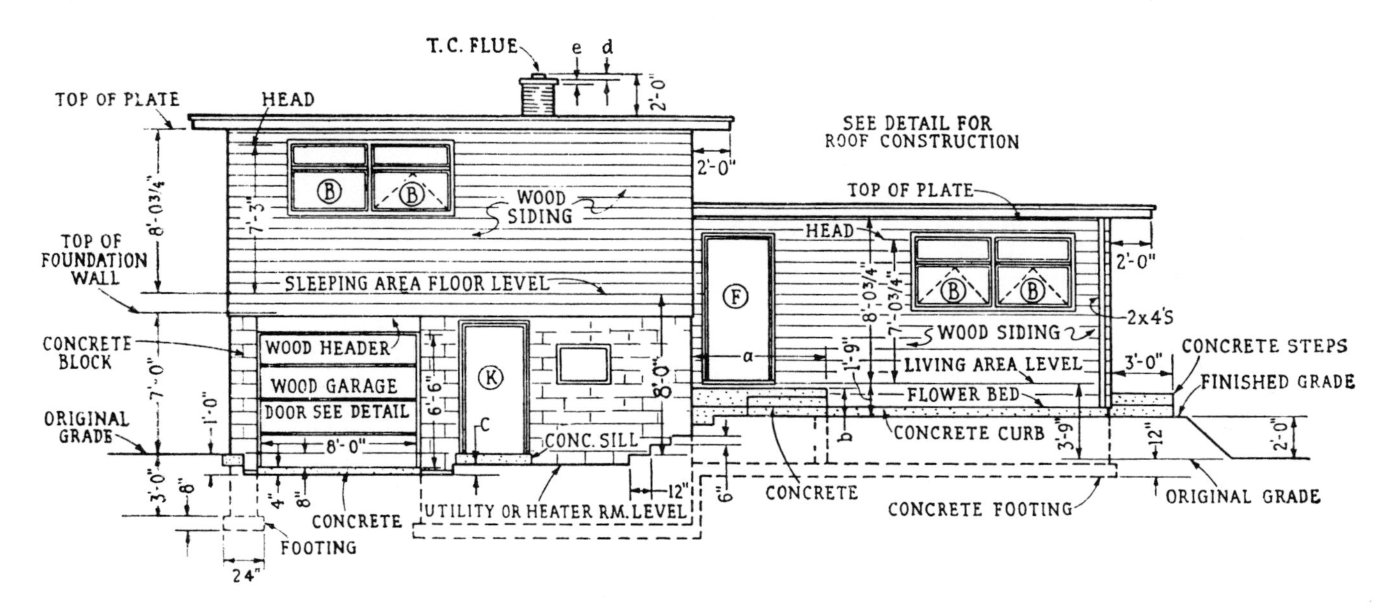

FRONT ELEVATION

1/4" = 1'-0"

0 1 2 3 4 5 6 7 8 9 10 11 12 13 14

REDUCED SCALE

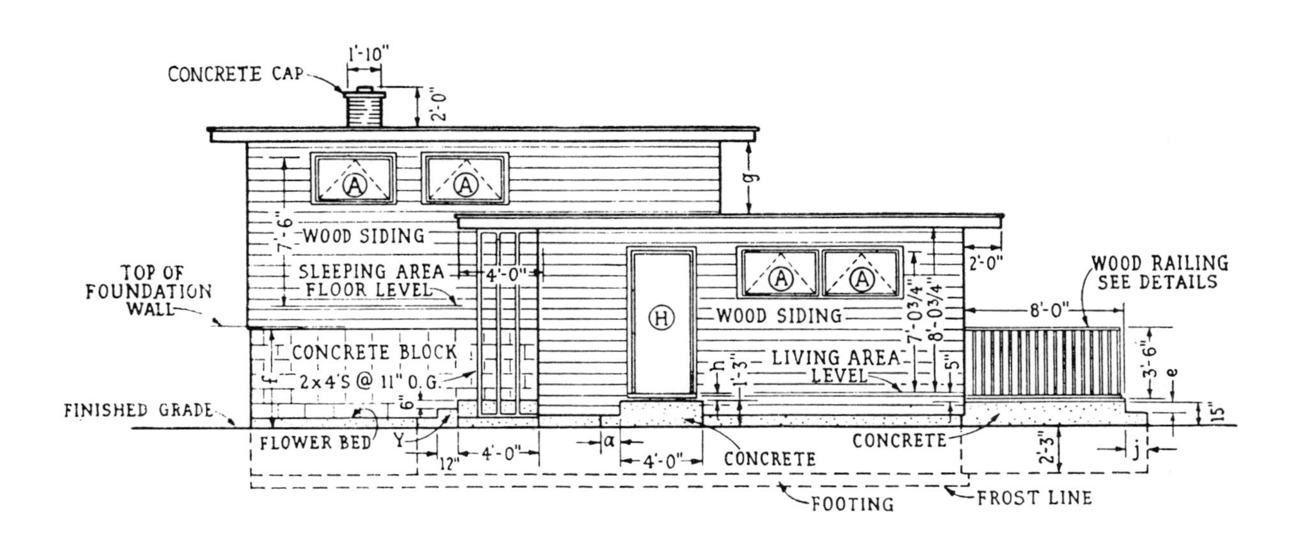

RIGHT SIDE ELEVATION

1/4" = 1'-0"

0 1 2 3 4 5 6 7 8 9 10 11 13 14 15

REDUCED SCALE

Figure 11

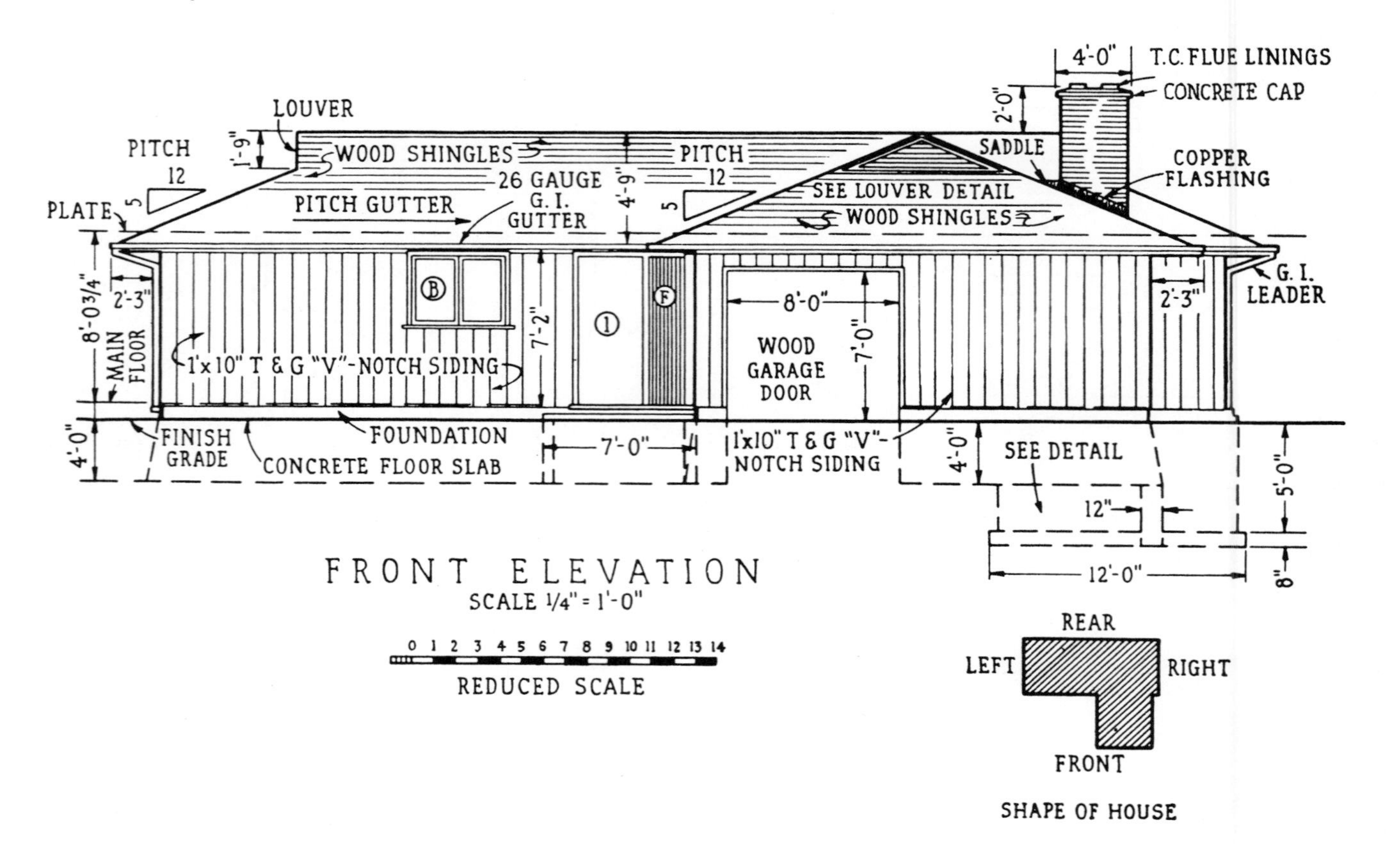

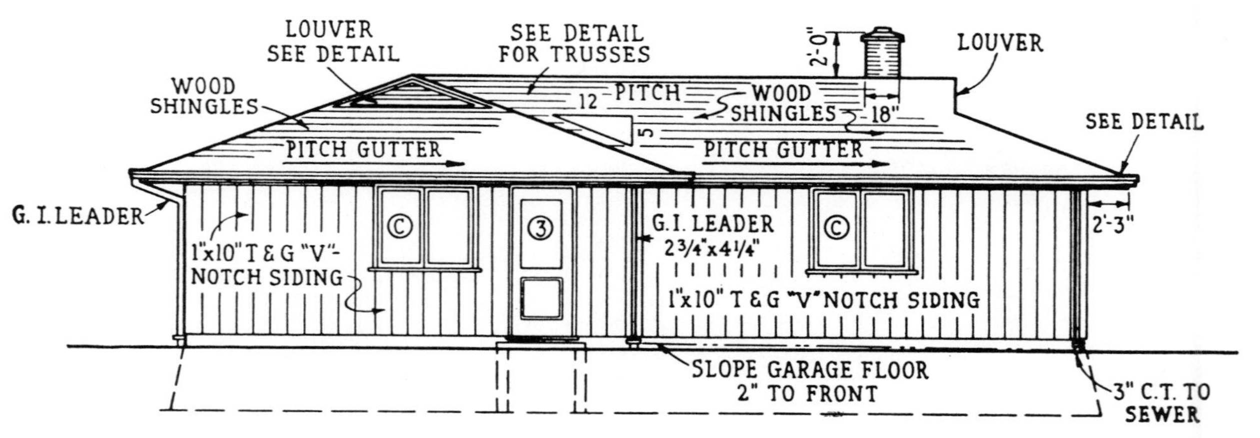

LEFT SIDE ELEVATION

SCALE 1/4"=1'-0"

0 1 2 3 4 5 6 7 8 9 10 11 12 13 14

REDUCED SCALE

Figure 12

CONSTRUCTION-PLAN STUDIES

Figures 11, 12, and 13 represent elevation views, as designers prepare them and as builders use them. In this book, because of page-size limitations, the elevations have to be shown reduced in size. We can study these drawings just as though they were regular

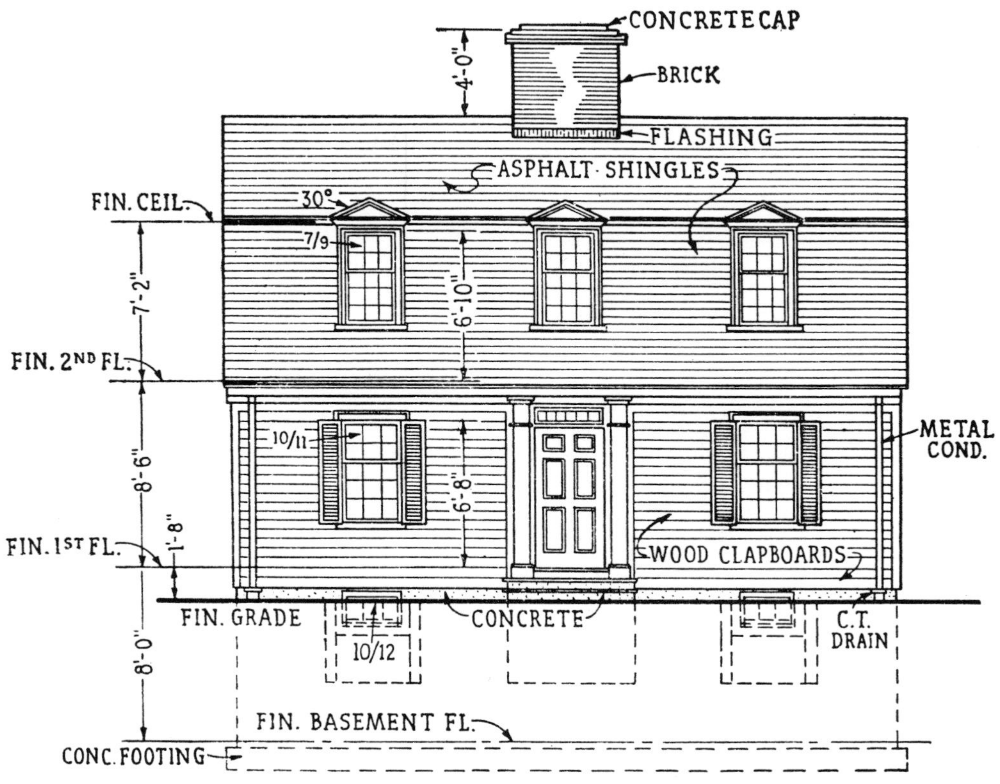

FRONT ELEVATION
SCALE 1/8"=1'-0"

NOTE:
ALL DETAILS MENTIONED ARE SHOWN IN DETAIL SHEETS.

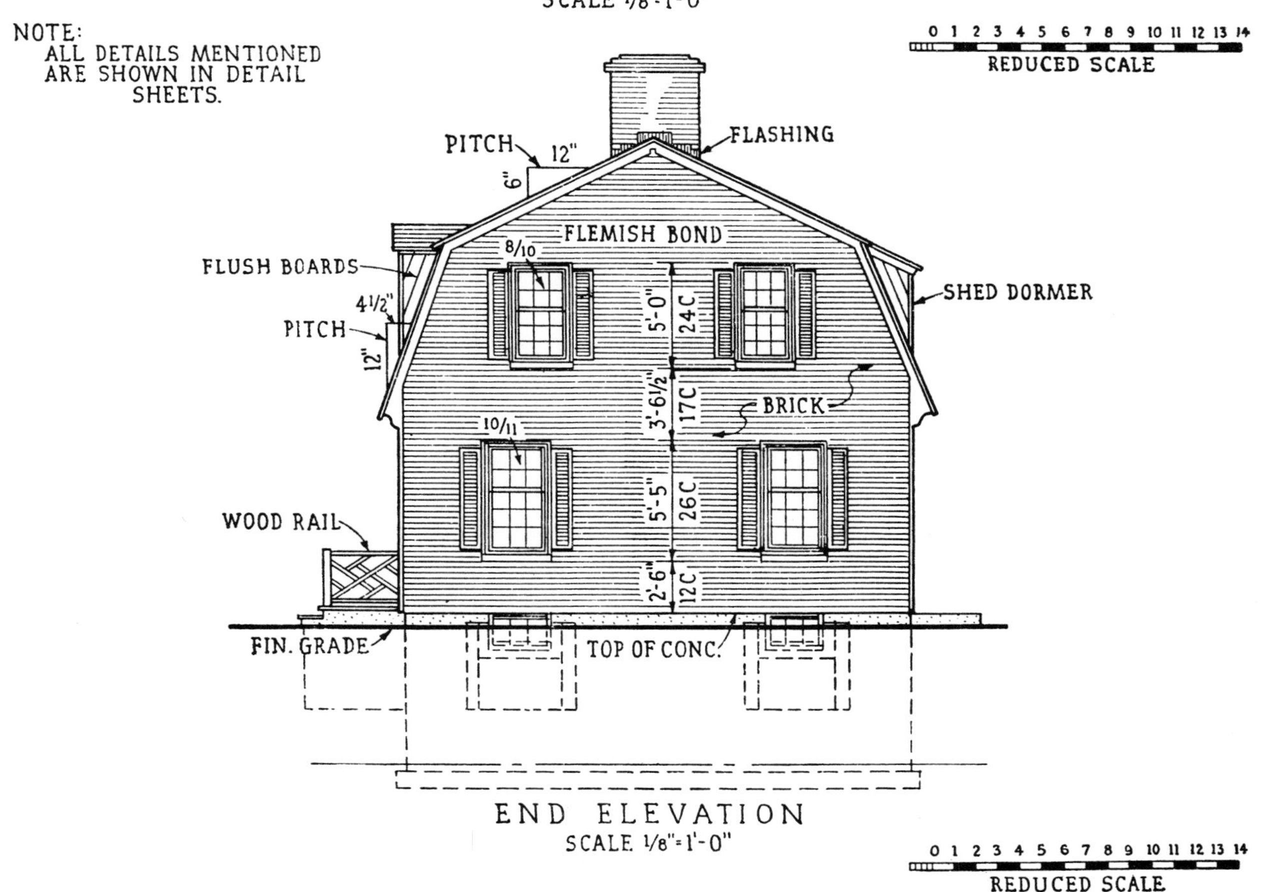

END ELEVATION
SCALE 1/8"=1'-0"

Figure 13

drawing size to learn what they intend to have built.

Lines The use of different weights of lines, on elevation views, aids our visualization. For example, the heaviest lines indicate the main outlines or shapes of the houses. Somewhat lighter lines are used to indicate the outlines of windows and doors. The lines used to show symbols are still lighter. Dimension lines are the lightest of all. Dashed lines are used to show foundations and footings below grade and for other items as will be noted in the following.

Figure 11. The following observations will be helpful in learning to read elevation views for a split-level house. It is suggested that readers locate all of the symbols and dimensions on the elevations, as each is pointed out in the following:

Materials The elevation view symbols show what various structural items are to be made of:

Chimney	*brick*	Foundation	*concrete block*
Chimney cap	*concrete*	Garage door	*wood*
Chimney flue	*T.C.*	Porches	*concrete*
Garage floor	*concrete*	Porch steps	*concrete*
Siding	*wood*	Flower-bed curbs	*concrete*
Railings	*wood*		
Footings	*concrete*		

Details In connection with the roof construction, porch railing, and garage door, there are notes on the drawings which refer to *details*. We shall study details in a later chapter.

Windows All regular window symbols show either letter *A* or *B*. The letters mean that the sizes of the windows are shown in a schedule which appears in connection with plan views and that the specs will explain other needed information. We shall study plan views and specs in later chapters. The dashed lines in the window symbols indicate that the windows open on hinges which are located where the dashed lines meet.

Doors The door symbols also have letters which mean that sizes will be given by a schedule in connection with plan views and that specs will give added and necessary information.

Roofs Flat roofs are indicated. When pitch is not indicated on the elevation views, the specs give the necessary information.

Levels The front elevation shows the three levels on which the house is to be constructed. The sleeping-area floor level is indicated just above door *K*. The living area is indicated just below the two windows marked *B* on the right side. The utility floor level is indicated to the right of the door marked *K*.

Utility-area Walls Both elevation views show that the walls of the utility area are to be made of concrete block.

Garage Floor The top of the garage floor is to be 8 inches below the top of the general floor level in the utility area.

General The sill for the utility room door *K* is to be made of concrete.

The finished grade is to be 2′ 0″ above the original grade.

Exterior steps are to be provided between the living- and utility-area levels, as indicated to the left of, and below, door *F*.

A concrete curb must be provided for flower beds 4′ 0″ wide on the right side of the utility-room wall and along the front of the living area.

The roof, along the front of the living area, must extend out 4′ 0″ and be supported at the right end by a group of three vertical 2 by 4s which are 11 inches on centers. The 2 by 4s are shown at the right-hand end of the front elevation and noted in the right-side elevation.

The concrete porch, outside of the *H* door in the right-side elevation, must have two steps. These steps are also shown at the right-hand end of the front elevation.

The railing and concrete symbols, shown at the right-hand end of the right-side elevation, indicate that there must be a back porch without a roof. The porch floor must be made of concrete and be two steps above finished grade. The dashed lines, under the porch, indicate that a foundation is required.

The dashed lines under the house show that foundations and footings are necessary.

Footings are concrete and must be 2′ 0″ wide, 8 inches thick.

The horizontal lines, which compose the siding symbol, indicate that the siding must be applied horizontally. Additional information concerning the siding, such as specific type of wood and nailing, will be found in the specifications.

The top of the T.C. flue lining in the chimney must be 2′ 0″ above the ridge of the roof.

The top of the foundation wall around the utility or heater area must be 7′ 0″ above original grade.

The garage floor must be 4 inches thick.

The rear porch railing must be 3′ 6″ high.

The risers for the steps outside of door *F* must be 6 inches high.

Position of Windows The heads of the *B* windows, in the front-elevation view of the sleeping area, must be 7′ 3″ above the sleeping-area floor level. The heads of the *B* windows, in the front-elevation view of the living area, must be 7′ 0¾″ above the living-area level.

The heads of the *A* windows, shown in the sleeping area in the right-side elevation, must be 7′ 6″ above the floor level. The heads of the *A* windows, in the living area, in the right-side elevation, must be 7′ 0¾″ above the floor level.

Concrete Porches The rear porch, shown at the right-hand end of the right-side elevation, must be 8′ 0″ wide and 15 inches thick. It must have a foundation 2′ 3″ deep.

The porch, in connection with door *H*, must be 3′ 0″ wide, as indicated by the front elevation, and 4′ 0″ long, as indicated by the right-side elevation.

Plate Heights The top of the wall plate for the sleeping area must be 8′ 0¾″ above the floor level of that area. The same dimension is shown for the top of the plate in the living area.

Roof Overhang All roofs, except over the door *F* and over the flower bed that is in front of the living area, must overhang a distance of 2′ 0″. Over the living-area flower bed, the overhang must be 4′ 0″.

Floor Levels As shown in the front elevation, the utility- or heater-room floor level must be at the level of the original grade. The sleeping-area floor level must be 8′ 0″ above the utility-room floor level. The living-area floor level must be 3′ 9″ above the original grade.

Missing Dimensions As indicated by the letters *a*, *b*, *c*, *d*, and *e*, in the front elevation, and by the letters *a*, *e*, *f*, *g*, *h*, and *j*, in the right-side elevation, many useful dimensions are not shown. However, as previously explained, they can be obtained by the scaling process. For example, missing dimension *a*, in the front elevation, is actually 6′ 6″.

Readers are urged to study the elevations in Figure 11 until they can visualize all of the details shown. Frequent reference to the perspective, in Figure 10, will be helpful.

Figure 12. The following information will also be helpful in learning how to read elevation views for a one-story house. Readers are urged to locate all symbols and dimensions on the elevations, as each observation is pointed out in the following:

Materials The elevation view symbols show what various structural details are to be made of:

Chimney	*brick*	Shingles	*wood*
Chimney flues	*T.C.*	Garage door	*wood*
Chimney cap	*concrete*	Flashing	*copper*
Leaders	*G.I.*	Gutters	*G.I.*
Floor slab	*concrete*	Louver	*wood*
Saddle	*copper*	Siding	*wood*

Details In connection with the louvers and some other items, there are notes which refer to detail drawings. As previously explained, we shall study such details in a later chapter.

Windows All regular window symbols show either letter *B*, *C*, or *F*. These letters refer to a schedule which we will study in connection with plan views.

Doors The door symbols contain numerals which also refer to schedules appearing on plan views.

Level All rooms are on one level.

Roof All roof areas are to have a pitch of 5 inches in 12 inches.

Louvers As shown, there are to be louvers at all gables.

General The leaders are to empty into 3-inch-crock draintile at grade level.

The long arrows, along the cornice, indicate the direction of pitch for the gutters.

The garage door opening must be 7′ 0″ by 8′ 0″. The garage door is wood. Additional information, such as type of door and opening mechanism, will be found in the specifications.

The under side of the cornice must be 7′ 2″ above the top of the foundation.

Roof overhang must be 2′ 3″.

The top of the wall plate must be 8′ 0¾″ above the top of the foundation.

Dimensions In these elevation views, the extent of the dimension lines is indicated by arrows.

Window Positions No vertical dimensions are necessary because the window heads are to be right under the bottom surface of the cornice.

Chimney The top of the chimney must be 2′ 0″ above the ridge of the roof. The chimney must be 4′ 0″ wide and 18 inches thick.

The chimney must have two flues and a concrete cap.

Scaling Any distances not shown by dimensions can be scaled as previously explained.

Figure 13. We shall use nearly the same procedures, as followed with respect to Figures 11 and 12, in observing what the symbols in this illustration indicate.

Materials The elevation-view symbols show what the various structural details are to be made of:

Front walls	*clapboard*	Chimney	*brick*
End walls	*brick*	Chimney cap	*concrete*
Roof	*asphalt shingles*	Leaders	*metal*
		Leader drains	*C.T.*
Foundation	*concrete*	Entrance steps	*concrete*
Footings	*concrete*	Dormer sides	*wood*
Railing	*wood*	Entrance	*wood*

Windows In the front elevation, the three second-floor window heads must be 6′ 10″ above the second-floor level and each window must contain fifteen 7- by 9-inch panes of glass. The two first-floor window heads must be 6′ 8″ above the first-floor level and each window must contain fifteen 10- by 11-inch panes of glass.

In the end elevation, the heads of the two first-floor windows must be 2′ 6″ + 5′ 5″, or 7′ 11″ above the top of the foundation and each window must contain fifteen 10- by 11-inch panes of glass. The heads of the two second-floor windows must be 3′ 6½″ + 5′ 0″, or 8′ 6½″ above the heads of the first-floor windows and each window must contain fifteen 8- by 10-inch panes of glass.

The notations, such as 12*C* and 17*C*, indicate the number of brick courses between the dimension symbols.

Doors Only one door symbol is shown and the note indicates that its construction is shown in a detail drawing.

Level The elevations indicate a two-story house. Or, in other words, there is to be a first and second floor.

Roof The roof must have two different pitches. The pitch symbol for the lower part of the roof indicates a slope of 12 inches in every 4½ inches. The pitch symbol for the upper part of the roof indicates a slope of 6 inches in every 12 inches.

General As indicated in the front elevation, there must be a basement which has a floor level 8′ 0″ below the level of the first floor.

There must be two basement windows in each elevation and each window must contain three 10- by 12-inch panes of glass.

Across the back of the house, at the second-floor level, there must be a shed-type dormer.

In the front elevation, only the first-floor windows show shutters. In the side elevation, both first- and second-floor windows indicate shutters.

Dimensions As in Figure 12, the extent of all dimension lines is indicated by arrows.

Chimney The cap of the chimney must be 4′ 0″ above the ridge of the roof.

Floor Levels The first-floor level must be 1′ 8″ above the finished grade, the second-floor level 8′ 6″ above the first-floor level, and the second-floor ceiling 7′ 2″ above the second-floor level.

Readers are again urged to study the elevation views discussed in this chapter until each symbol is understood and the views are visualized.

QUESTIONS AND ANSWERS

The following questions and answers concern Figures 14 and 15, which show two elevation views for a sloping-roof, brick and shingle house. As a means of making sure that you have learned to visualize and read such important aspects of construction plans, answer the questions orally or in written form and then check your reading with the *descriptive* answers shown. A review of this sort provides helpful practice and increases your ability to read drawings.

Question 1 Is the represented house of the one-story, two-story, or split-level type?
Answer 1 The various levels at which window symbols are shown indicate the house is of the split-level type.

Question 2 How many levels are indicated?
Answer 2 The front elevation shows three levels. At the left-hand side of the drawing a living-room and kitchen level is indicated. At the right-hand side of the same view, bedroom and garage and den levels are noted.

Question 3 Which of the three levels is the highest?
Answer 3 Both of the elevation views show that the bedroom level is the highest. The front elevation shows that the garage and den floor level, or basement level, is the lowest. At the left-hand side of the rear elevation, there are dimensions which indicate the heights of the three floor levels. The living-room and kitchen level is 4′ 3″ above the garage and den level, and the bedroom level is 3′ 9″ above the living-room and kitchen level.

Question 4 What ceiling height is required for the living-room and kitchen level?
Answer 4 At the left-hand side of the front elevation and at the right side of the rear elevation, a dimension indicates that the ceiling height must be 8′ 1″.

Question 5 Which of the three levels has the least ceiling height?
Answer 5 At the right-hand side of the front elevation there are 7′ 6″ and 8′ 1″ dimensions which show ceiling heights for the garage and den and bedroom levels. The bedroom ceiling height is the same as for the living-room and kitchen level. Thus, the garage and den level is to have the least ceiling height.

Question 6 How many louvers are shown in the two elevation views?
Answer 6 Both the front and rear gable ends, over the bedroom level, include louver symbols. Thus, two louvers are shown.

Question 7 Of what materials are the louvers to be made?
Answer 7 There are no notes, or other indications, referring to details. Thus, we can conclude that the louvers are to be made of wood and that, as indicated in the symbols, screening is required over the openings.

Question 8 Where is flashing indicated on the elevations and what material is necessary?
Answer 8 The front elevation shows that copper flashing is required between the chimney and the roof.

Question 9 What do the letters in the window symbols indicate?

Answer 9 The letters indicate that the window sizes and other needed information will be given in schedules which appear on other drawings, and in the specs.

Question 10 What kind of exterior finish is required for the walls?
Answer 10 Three types of exterior finish are indicated. The front wall of the living and kitchen area is to be brick veneer. That is indicated by symbol and by the note which appears under the symbol for the *C* windows. Both elevations indicate that wood siding is required for the gable ends. Both elevations also indicate that wood shingles are to be used below the gable ends.

Question 11 Is there to be a basement under the whole house?
Answer 11 The dashed lines, on the bottom of the front elevation, indicate that foundations and footings are required. The "step down" note shows that parts of the foundation and footings are lower than other parts. This means that the area under the living-room and kitchen level is not excavated. The "crawl" note also indicates that the area is not to be excavated. In other words, the garage and den level constitutes the only basement area.

Question 12 What are the foundation dimensions?
Answer 12 The front elevation shows that foundations are to be 8 inches thick. Their depth, including the thickness of footings, is to be 4′ 0″.

Question 13 What footing dimensions are required?
Answer 13 At the left-hand side of the front elevation, there are dimensions which show that the footings must be 16 inches wide and 8 inches deep.

Question 14 Of what material are the foundations to be made?
Answer 14 The dotted symbol shows that concrete is to be used for the foundations. The "Conc." abbreviation also indicates concrete.

Question 15 Are shutters required?
Answer 15 Shutter symbols are shown for the *A* and *B* windows and on only the front elevation.

Question 16 What size garage door is indicated?
Answer 16 The garage-door symbol is shown on the front elevation. The dimensions given with it show that the door is to be 7′ 0″ by 8′ 0″.

Question 17 What materials are to be used in making the garage door?
Answer 17 Wood and glass. The "GL." abbreviation indicates that four panes of glass are to be built into the top portion of the door.

Question 18 How many individual panes of glass are required for the *A* windows?
Answer 18 The horizontal lines in the symbols represent muntins. Thus, each window must have four individual panes of glass.

Question 19 How far above the bedroom-floor are the heads of the *A* windows?
Answer 19 Near the *A* window, at the right-hand side of the front elevation, the distance is shown to be 6′ 9″.

Question 20 What materials are required for the chimney?
Answer 20 The main portion is to be made of brick. The flue must be lined with T.C. The cap must be made of concrete.

Question 21 What pitch is required for the roof?
Answer 21 The pitch symbol is shown on the front elevation. It indicates that the pitch, or slope, must be 6 inches for every foot of run, or length, of the roof.

Question 22 What material is required for the front entrance steps?
Answer 22 The dotted symbol indicates concrete.

Question 23 How many risers are required?
Answer 23 The symbol shows four horizontal lines. The bottom line represents the grade. The top line represents the surface of the porch floor outside the main entrance door. Thus, there are three risers.

Question 24 Where is wrought iron required?
Answer 24 For a trellis in connection with the front entrance and porch.

Question 25 What kind of roof shingles are required?
Answer 25 Notes on both elevations specify asphalt.

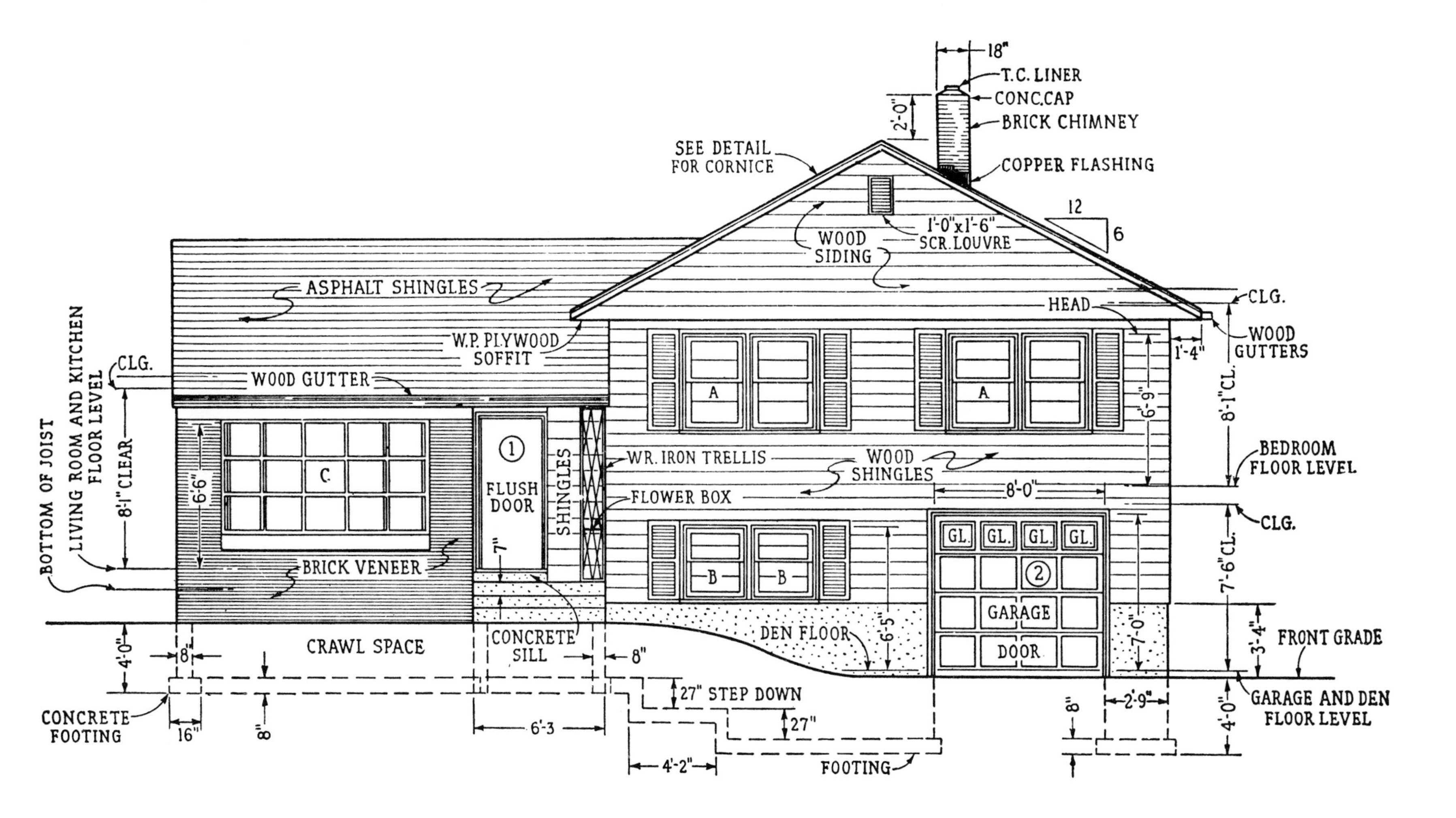

FRONT ELEVATION

1/4"=1'-0"

0 1 2 3 4 5 6 7 8 9 10 11 12 13 14

REDUCED SCALE

Figure 14

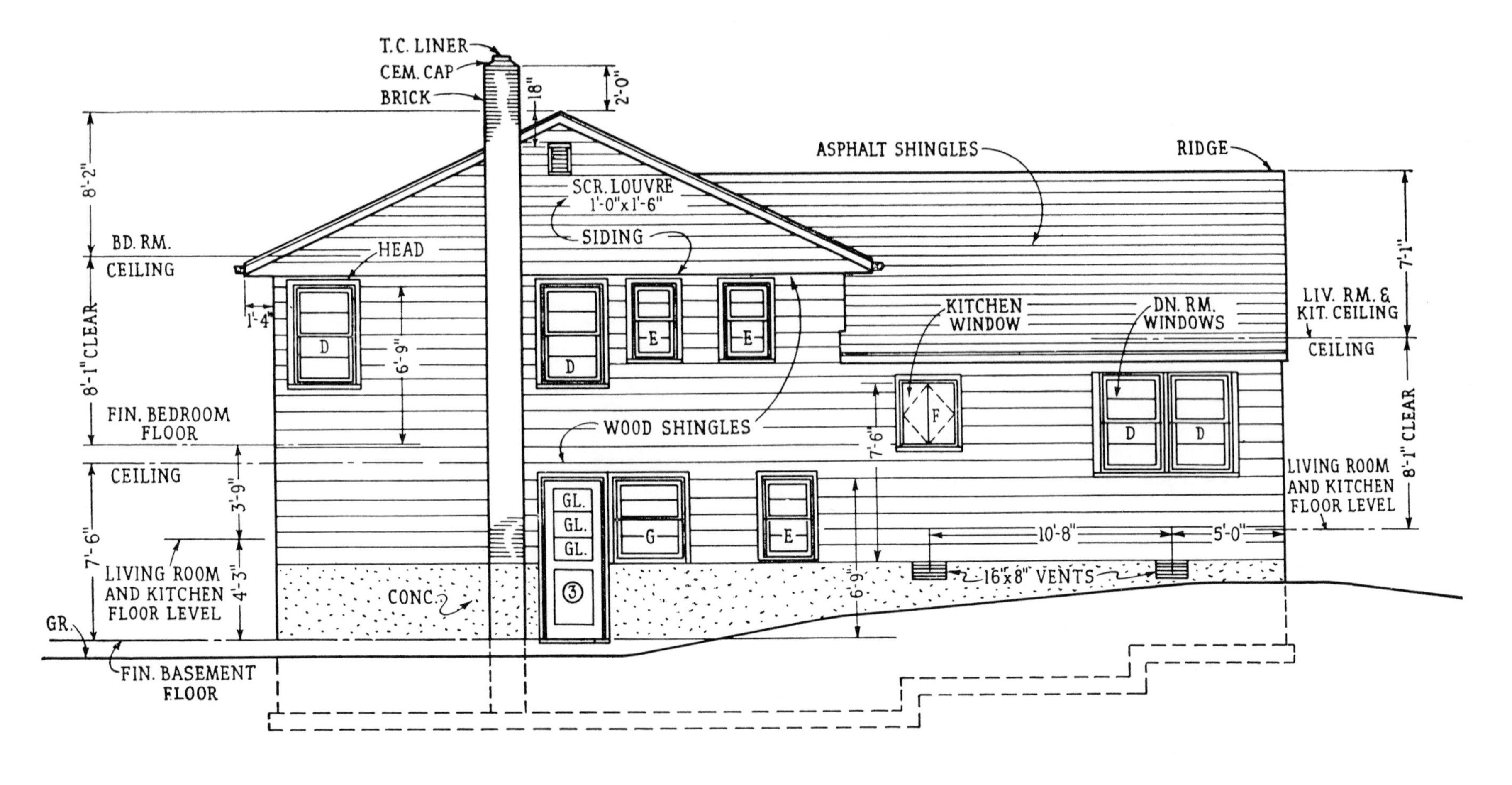

REAR ELEVATION

1/4"=1'-0"

0 1 2 3 4 5 6 7 8 9 10 11 12 13 14

REDUCED SCALE

Figure 15

Question 26 Where is plywood required?
Answer 26 On the front elevation, there is a note which points out that the under sides of the overhanging cornices are to be sheathed with plywood.

Question 27 Are any casement windows required?
Answer 27 The rear elevation shows a casement symbol in connection with the kitchen area. The dashed lines indicate that both parts of the window swing on hinges which are to be located at the point where the dashed lines meet.

Question 28 How many crawl-space vents are required?
Answer 28 The rear elevation shows two vent symbols.

Question 29 How high above the living-room and kitchen ceiling level is the ridge of the roof to be?
Answer 29 At the right-hand side of the rear elevation, the distance is shown to be 7′ 1″.

Question 30 Is the chimney to be within the walls of the house?
Answer 30 The rear elevation shows that the chimney is to be outside the wall.

Question 31 How high above the bedroom floor level is the ridge of the roof to be?
Answer 31 At the left-hand side of the rear elevation, the distance is shown to be 8′ 1″ + 8′ 2″, or 16′ 3″.

Question 32 How far is the head of window *E* to be above floor level?
Answer 32 The rear elevation shows the distance to be 6′ 9″.

Question 33 Which of the indicated windows are to be double hung?
Answer 33 The symbols for all windows, except windows *C* and *F*, indicate double-hung specifications.

Question 34 How far above grade is the level of the front porch?
Answer 34 The risers shown on the front elevation are specified as 7 inches high. The porch is 21 inches above grade.

Question 35 How wide is the front porch to be?
Answer 35 The front elevation shows a 6′ 3″ dimension.

Readers are urged to continue their study of Figures 14 and 15 until all the symbols are understood and until the house can be visualized from the two sides shown.

CHAPTER FOUR

Plan Views

In Chapter 3, we found that elevation views are picturelike representations of the *exteriors* of structures which are drawn to small scale and which include many symbols, terms, and abbreviations to indicate the various materials and structural parts. We learned to visualize the intended appearances and specified materials for the exteriors of proposed structures by studying the representative drawings. In other words, we learned how to read elevation view drawings.

Plan views are also picturelike representations in which symbols, terms, and abbreviations are used to indicate various materials and structural parts. However, plan views show the intended appearances and specified materials for the *interiors* of structures and employ symbols, terms, and abbreviations to a greater extent than required for elevation views. In order to read plan views, we shall have to learn more about visualization and become acquainted with many new symbols and terms.

In this chapter, we shall learn how to visualize plan views and the relationship between them and elevation views. We shall also learn the new symbols, terms, and abbreviations necessary. Thus, the purpose of this chapter is to teach us how to read plan views, and to advance our ability to read drawings to the point where we can visualize both the exteriors and the interiors of proposed structures.

HOW TO VISUALIZE PLAN VIEWS

The term *plan* is generally spoken of in two ways. All the drawings for a proposed structure are known as *plans.* More accurately, a plan is a view which shows how a structure appears from directly above. For a house, a plan shows the

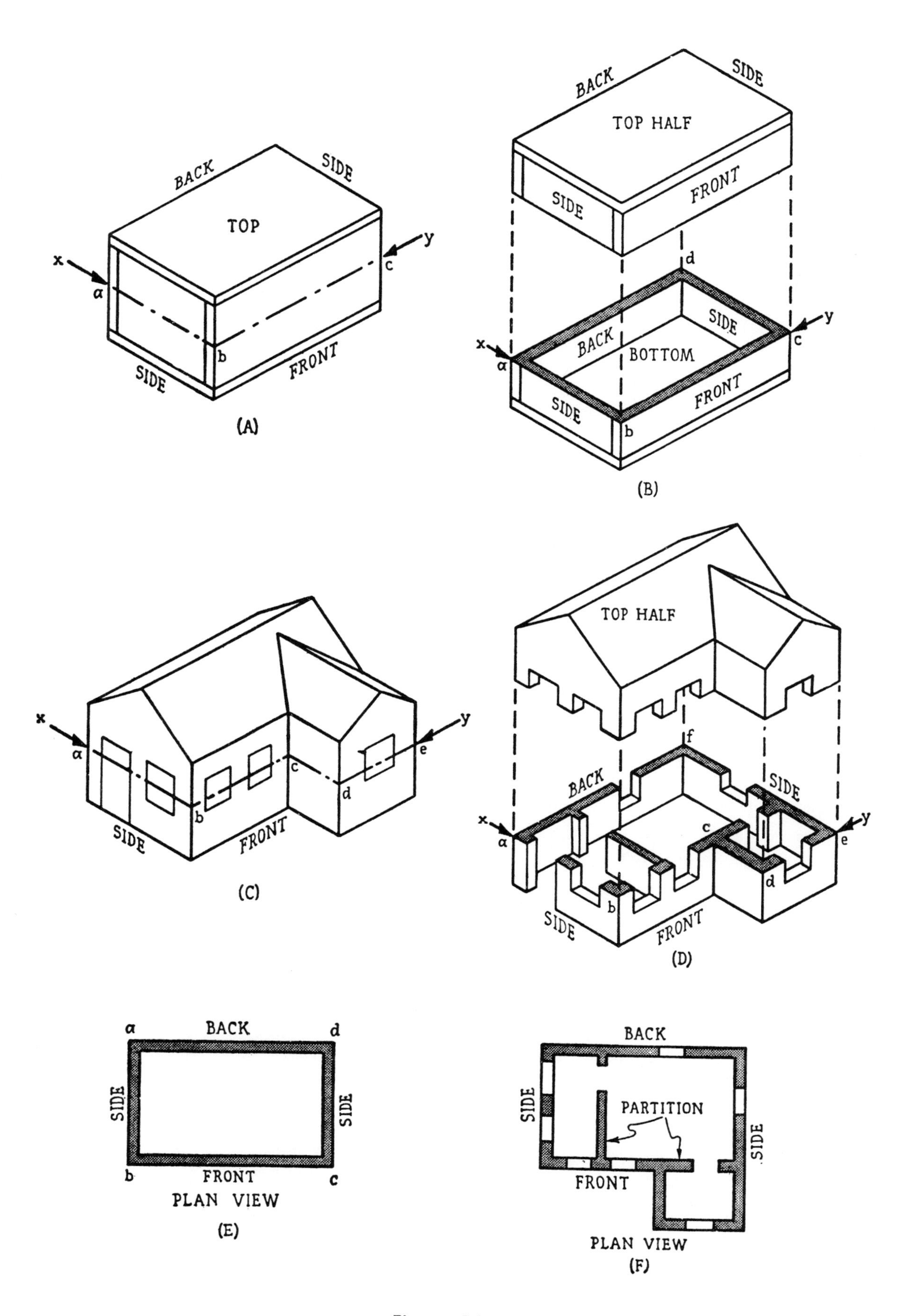
BACK
SIDE
TOP
x
y
a
b
c
SIDE
FRONT
(A)
BACK
SIDE
TOP HALF
SIDE
FRONT
d
x
y
BACK
BOTTOM
SIDE
a
c
SIDE
FRONT
b
(B)
x
y
a
b
c
d
e
SIDE
FRONT
(C)
TOP HALF
f
BACK
SIDE
x
y
a
c
e
d
b
SIDE
FRONT
(D)
a
BACK
d
SIDE
SIDE
b
FRONT
c
PLAN VIEW
(E)
BACK
SIDE
PARTITION
SIDE
FRONT
PLAN VIEW
(F)

Figure 16

layout or arrangement of rooms, and other parts, of the interior. In this connection, such views are called either *plan views* or *floor plans.* We shall use the term *plan view.*

We have learned that elevation views are actually *vertical* views. Plan views, on the other hand, are *horizontal* views which include the following typical information:

1. Arrangement, shapes, and sizes of all rooms or other areas.
2. Overall shape and size of the house, including the roofs.
3. Information about all materials.
4. Thicknesses of walls and other structural parts.
5. Locations of all the windows and doors.
6. Sizes, kinds, and locations of all items such as stairs, chimneys, fireplaces, closets, and porches.
7. Sizes, shapes, and locations of all bathroom and kitchen equipment.
8. Miscellaneous information which serves as a guide, or instructions, to the builders.

In order to visualize elevation views, we imagined that we could look at the sides of the house, as explained in connection with Figures 4 and 5 of Chapter 3. In order to visualize plan views, it is necessary for us to imagine that an existing structure can be cut in half horizontally and that the top part can be removed so that we can look straight down at the cut surfaces of the remaining part. What we would then see constitutes a plan view.

Examples The *A* part of Figure 16 shows a picturelike view of an ordinary wood box which has four sides, a top, and a bottom. It is similar to the object shown in the *A* part of Figure 4. Let us imagine that a large saw can be used to cut through the whole box. The dashed line, from *x* to *y*, indicates that the sawing will cut the box at *a*, *b*, and *c*. Next, let us imagine, as shown in the *B* part of the illustration, that after the sawing is done, the top half of the box can be moved upward far enough so that we can look directly down at the cut surfaces marked *abcd*. The heavy black cut surfaces of the front, back, and sides constitute a plan view. The *E* part of the illustration shows a typical plan view of the box. Note that the plan view shows all four sides. In other words, the plan view shows the walls of the box and their relationship to each other.

In the *C* part of Figure 16, indications have been added for room, windows, doors, and an ell, to make the box somewhat resemble a house. (In Figure 4, of Chapter 3, we also made an object look something like a house.) Let us imagine that the structure can be cut through as indicated by the dashed line from *x* to *y*. As before, we imagine that a large saw can be used for this purpose. Next, as with the *A* part of this illustration, let us imagine that the top half can be raised up so we can look directly down at the cut surfaces marked *abcdef* as shown at *D*. The cut surfaces, shown in heavy black, constitute a plan view of the structure. Note that the black cut surfaces indicate the door, the windows, and the rooms inside of the structure. The *F* part of the illustration shows a typical plan view. Note that the door, the windows, and the rooms, in addition to exterior and interior walls, are all shown. From such a plan view it is possible to visualize a great deal about the structure. In other words, we can discover how many windows are required, how many doors are needed, and how many rooms there are, and obtain a fairly good mental picture of the structure.

From our observations on the *C* and *D* parts of Figure 16, we learned that plan views are horizontal cuts, sometimes known as *sections*, which are imagined as being at such a level that they go through the windows and doors. Plan views are always drawn in this manner so as to show the positions of windows and doors in relation to walls and other structural parts.

The foregoing examples obviously concern a one-story house of the type shown by Figure 12 of Chapter 3. Some houses, such as the one

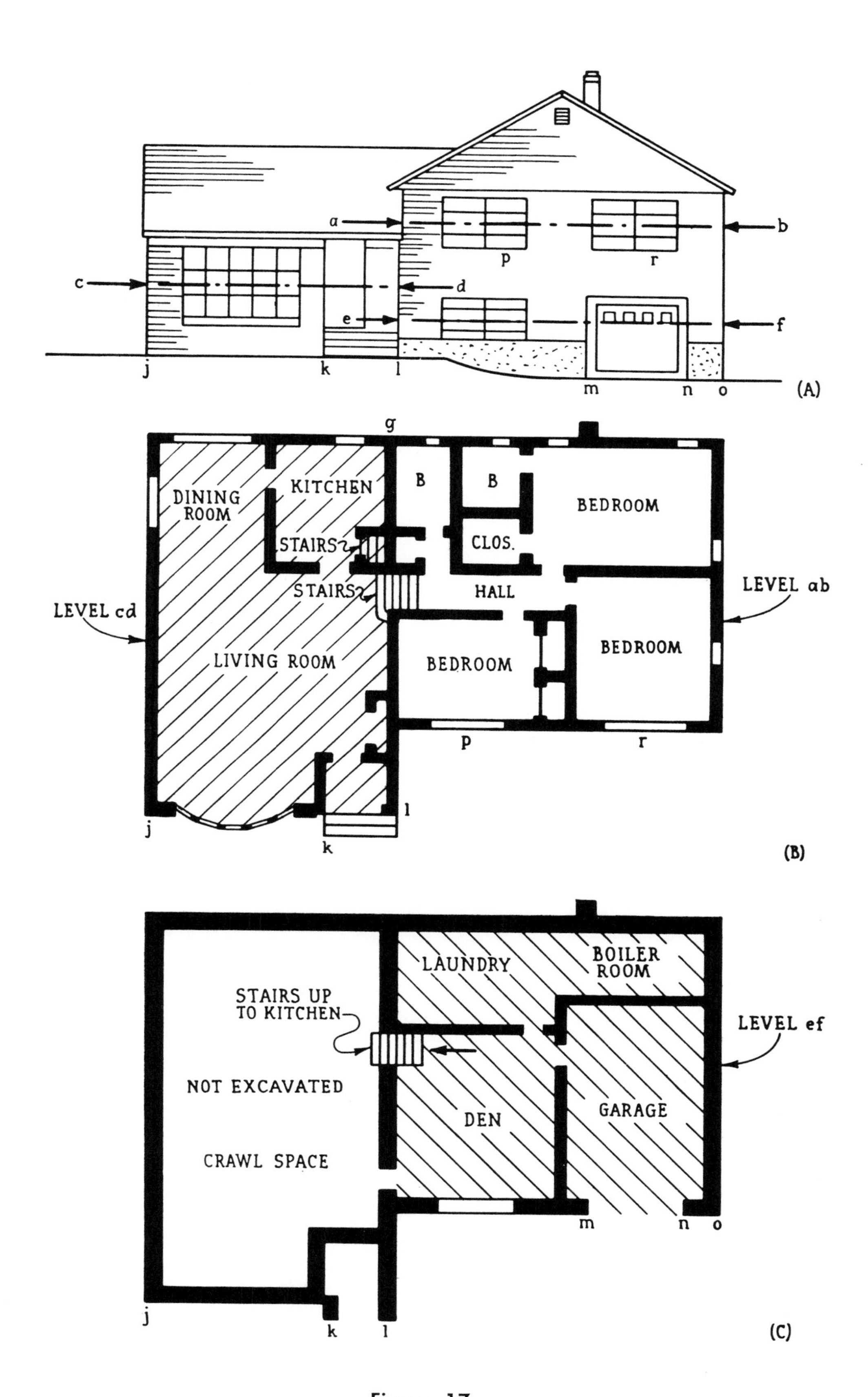
a
b
c
d
e
f
p
r
j
k
l
m
n
o
(A)
g
DINING ROOM
KITCHEN
B
B
BEDROOM
STAIRS
CLOS.
STAIRS
HALL
LEVEL cd
LEVEL ab
LIVING ROOM
BEDROOM
BEDROOM
p
r
j
l
k
(B)
LAUNDRY
BOILER ROOM
STAIRS UP TO KITCHEN
LEVEL ef
NOT EXCAVATED
DEN
GARAGE
CRAWL SPACE
m
n
o
j
k
l
(C)

Figure 17

shown by Figure 13 of Chapter 3, are of the two-story type and have basements. In such a case, plan views are required for both the first and second floors and for the basement. The plan views can be visualized by imagining that a house of this type could be cut through in three places: through the basement windows, through the first-story windows, and through the second-story windows.

Now that we know how to visualize the relationship of plan views to a structure, we can reverse the process when we see plan views. In other words, by studying any plan view, we can visualize the represented structure, especially if we have elevation views for the same structure before us. The relationship between elevation and plan views will be explained in the next section of this chapter.

Visualizing Split-level Plan Views The *A* part of Figure 17 shows a sketch of an elevation view of the same house we visualized, via questions and answers, in Figures 14 and 15 of Chapter 3. Thus, we are already familiar with the *ab*, *cd*, and *ef* levels shown in Figure 17, which represent, respectively, the bedroom, the living room and kitchen, and the den and garage levels.

Note that in the *A* part of Figure 17 the dashed lines (similar to dashed lines *x* and *y* in Figure 16) cannot pass through more than one level and at the same time cut the windows and doors. For that reason, we must visualize that levels *ab* and *cd*, for example, are cut at the same time and that all parts of the structure above these levels can be raised up together. The *B* part of Figure 17 shows the cut surfaces. The wall marked *gl* is the division between levels. The living-room and kitchen level, shown shaded, is actually below the bedroom level. This fact is indicated by the stairs, which show that the bedroom level is five steps higher than the living-room and kitchen level. Even though these two levels cannot be cut by the same cut, so to speak, they are shown as though at the same level in order to bring out the relationship between them.

If we imagine that the structure is cut through at the *ef* level, as indicated in the *A* part of Figure 17, we will see the cut surfaces shown in the *C* part of the illustration. The fact that the den and garage level is lower than the living-room and kitchen level is indicated by the steps. In the *B* part of the illustration, note that the steps from the den and garage level are shown in the kitchen.

We can now realize that in a real house of this type and design, we would go down steps in going from the bedroom level to the living-room and kitchen level and down more steps in going from the kitchen to the den or garage.

Designers usually draw plan views so that the front of the structure, as indicated in Figures 16 and 17, is nearest to us as we look at the drawings representing it.

RELATIONSHIP BETWEEN PLAN AND ELEVATION VIEWS

From our study of elevation views we learned that their purpose is to show the exterior of a proposed structure after all structural work is complete. Such views indicate the kinds and approximate positions of all windows, doors, chimneys, porches, and everything else which can be seen from the exterior. Such information may also give us some ideas concerning the interior arrangement of rooms. For example, a picture window often indicates one side or end of a living room. A door may suggest one wall or one side of a kitchen. A porch may suggest another room. However, unless the elevation views are very similar to another structure with which we are well acquainted, both inside and out, we cannot very accurately guess what the interior of the structure is like merely from a study of the elevation views.

From our study of Figures 16 and 17 we learned how to visualize plan views and that they show what the interior arrangement of the structure is to be, so far as sizes and shapes of rooms, locations of windows and doors, etc.,

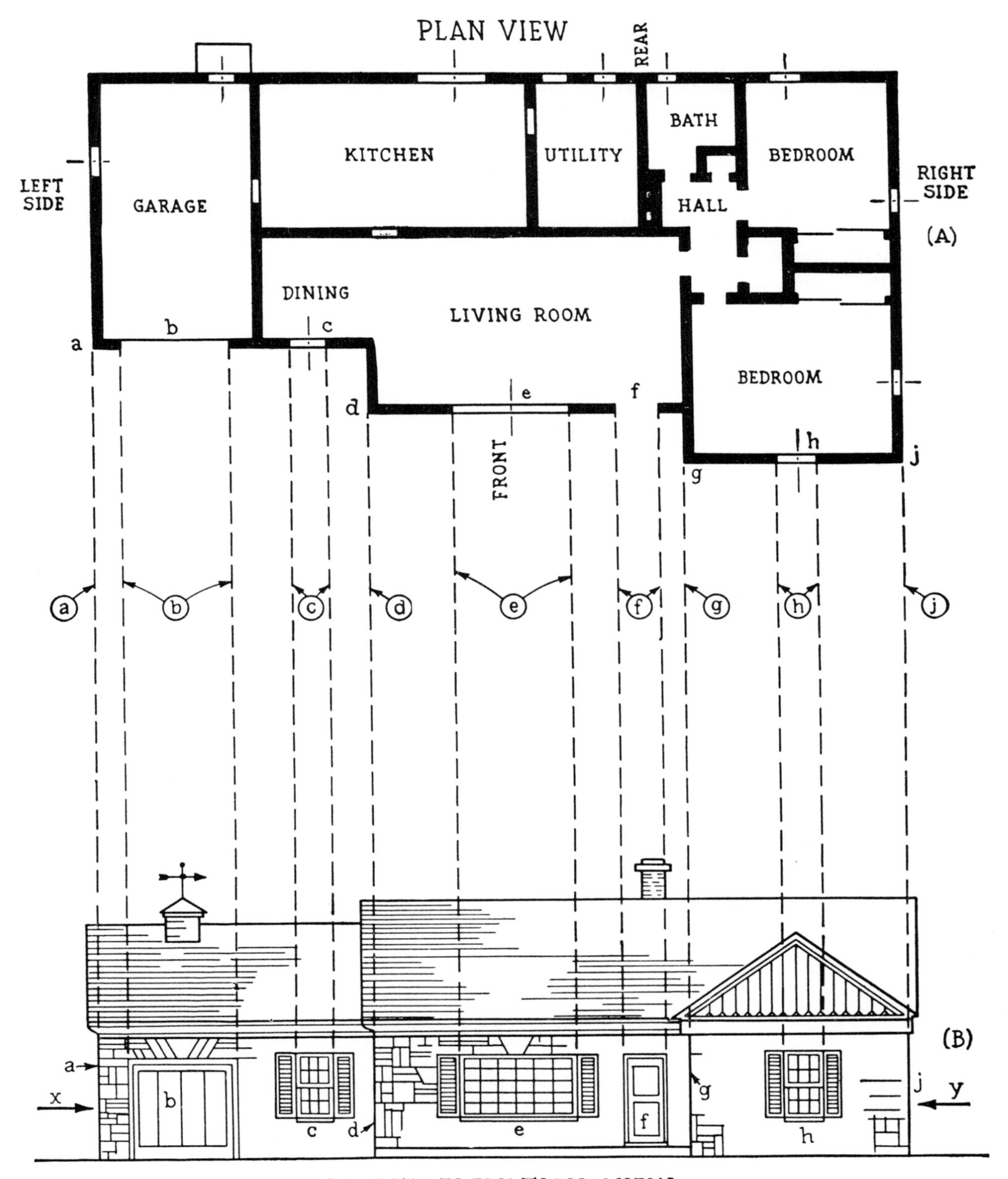
PLAN VIEW
REAR
KITCHEN
UTILITY
BATH
BEDROOM
LEFT SIDE
GARAGE
HALL
RIGHT SIDE
(A)
DINING
LIVING ROOM
BEDROOM
FRONT
(B)
x
y
FRONT ELEVATION VIEW

Figure 18

are concerned. Later in this chapter, we will learn a great deal more about plan views. Here, however, we are interested only in the visual relationships between plan views, as we know them at this stage of our study of drawing reading, and elevation views, with which we are now thoroughly acquainted. It is best that we understand this relationship before going on with our study of plan views.

In the *A* part of Figure 18, the heavy black lines can be thought of as indicating a cut surface as in Figures 16 and 17. In other words, we can imagine that the structure shown in the *B* part of Figure 18 has been cut through, from *x* to *y*, and that the top part, including the roof, can be moved up so we can see the plan view shown in the *A* part of the same illustration.

If we looked at only the elevation view of a structure as shown in Figure 18, we might form the opinion that the window at *e* was part of a living room, that maybe the door shown at *f* opened into a hall, and that the door at *b* opened into a garage. We cannot even guess in what sort of rooms the *c* and *h* windows might be. If we had all four elevation views, for the same structure, our guess about the interior would be just as difficult and fruitless.

By looking at only a plan view of a house, such as that in Figure 18, we might form a little better idea of the whole structure, both inside and out, than we could from the elevation views alone. However, the plan view in turn shows very little information about the exterior appearance.

Thus, we can realize that both plan and elevation views are necessary if we hope to visualize a structure correctly from both the inside and the outside.

There is a definite and accurate relationship between plan and elevation views, as shown graphically in Figure 18. For example, note the corners *a* and *j* in both the elevation and plan views. The dashed lines, also marked *a* and *j*, show how these two corners are related so far as the two views are concerned. Next, note the garage door which is marked *b* in both of the views. The dashed lines, also marked *b*, indicate the relationship between the two views. In like manner, the windows, corners, and door, marked *c*, *d*, *e*, *f*, *g*, and *h* in both views, indicate the relationship between the views.

If the rear and both side elevations for the structure shown in Figure 18 could be projected, like the front elevation, complete visualization of the structure would be easy. But construction plans are not made in this manner. To aid visualization, however, we can hold the various elevation views in front, in back, or to either side of a plan view as suggested in Figure 18. If we have all the prints for a proposed structure, and hold the elevation views as suggested, we can soon learn to visualize the whole structure by looking at the various views separately.

The relationship between plan and elevation views for split-level is just as easy to visualize, if we project imaginary lines between the two views as shown in Figure 18.

Note the points *j*, *k*, *l*, *p*, and *r* in the *B* part of Figure 17. These same points are indicated, using the same letters, in the elevation view shown in the *A* part of the same illustration. In like manner, the points *j*, *k*, *l*, *m*, *n*, and *o* in the *C* part of the illustration are also shown in the *A* part.

For two-story houses, the relationship between plan and elevation views is visualized in the same manner. The relationship between basement plan views and elevation views is visualized as explained for Figure 18 and as noted for the points *m*, *n*, and *o* in the *A* and *C* parts of Figure 17.

SYMBOLS

Earlier in this chapter we learned how to visualize plan views by imagining that a structure could be cut horizontally in half (see *C* and *D* parts of Figure 16) and that the part above the cutting line *xy* could be moved away, so that we could look down at the cut surface of the part below the line. In other words, when we think about or look at plan views, we imagine that we are directly above them and that we are looking straight down at them.

When we are directly above any object and look straight down at it, the picture we see has a *flat* appearance because the sides, or vertical areas, are not visible. For example, if we could look straight down at the object shown in the *A* part of Figure 4, we would see a flat area which is rectangular. In like manner, when we look at plan views we also see flat areas, or shapes, as illustrated in the *E* and *F* parts of Figure 16.

Plan views are picturelike representations of the interiors of structures. However, the pictures shown in them do not look nearly as much like the structural parts represented as is the case with elevation views. We can easily understand this if we imagine that the *F* part of Figure 16 contains windows. What we would see of them is nothing like the window symbols shown in an elevation view. Thus, the "pictures" we see in plan views are much more *symbolic* than the "pictures" in elevation views.

Because plan views are *flat*, they require the use of more symbols than are ordinarily found in elevation views. For example, in an elevation view we can see the windows in their vertical positions. We recognize them as windows by seeing just one symbol—that of the window. On the other hand, *two* symbols are necessary in order to indicate a window in a plan view. First, a wall symbol is necessary. Then, the window symbol is shown within the wall symbol. In like manner, many other plan-view items each require the use of two symbols whereas the same items can be shown by one symbol each in elevation views.

Elevation-view symbols can be drawn without too much concern for scale. In other words, if they are drawn only fairly well to scale they serve their purpose without confusion or trouble. In plan views, every symbol (except those to indicate materials) must be carefully and accurately drawn to scale. We will understand the necessity for such accuracy a little later on when we study actual plan views.

SYMBOLS USED IN PLAN VIEWS

In plan views, unlike elevation views, symbols are the only means of showing the required information. Everything shown or indicated on plan views is accomplished by means of symbols. Thus, so far as plan views are concerned, these symbols constitute the language by which the views are described. Readers are urged to memorize all the symbols shown and explained in the following. Unless all symbols can be visualized and understood, the reading of plan views will be most difficult.

Material Symbols Figures 19 and 20 show all the commonly encountered material indications used in connection with plan views. Many of them are similar in appearance to the material symbols used for elevation views; yet there is a distinct difference in the way they are used.

Walls and Partitions Both of these items are shown in plan views by heavy parallel lines whose distance apart, strictly according to scale, indicates the thickness of the parts; thus the exterior walls and the interior walls (partitions) for a structure are indicated by parallel lines. In addition, a symbol is shown between the parallel lines to indicate what material is to be used. In Figures 16 and 17 a solid black symbol was used simply because we were not ready to consider specific symbols at that time.

If the exterior walls of a structure are to be made of wood (called *frame* walls), the space between the parallel lines would be left blank, as indicated by the symbol shown at *A* in Figure 19. On the other hand, if the walls are to be made of brick, the space between the lines would be filled with the symbol shown at *C*. If brick-veneer walls are required, the symbol shown at *N* would be placed between the

PLAN VIEW MATERIAL SYMBOLS

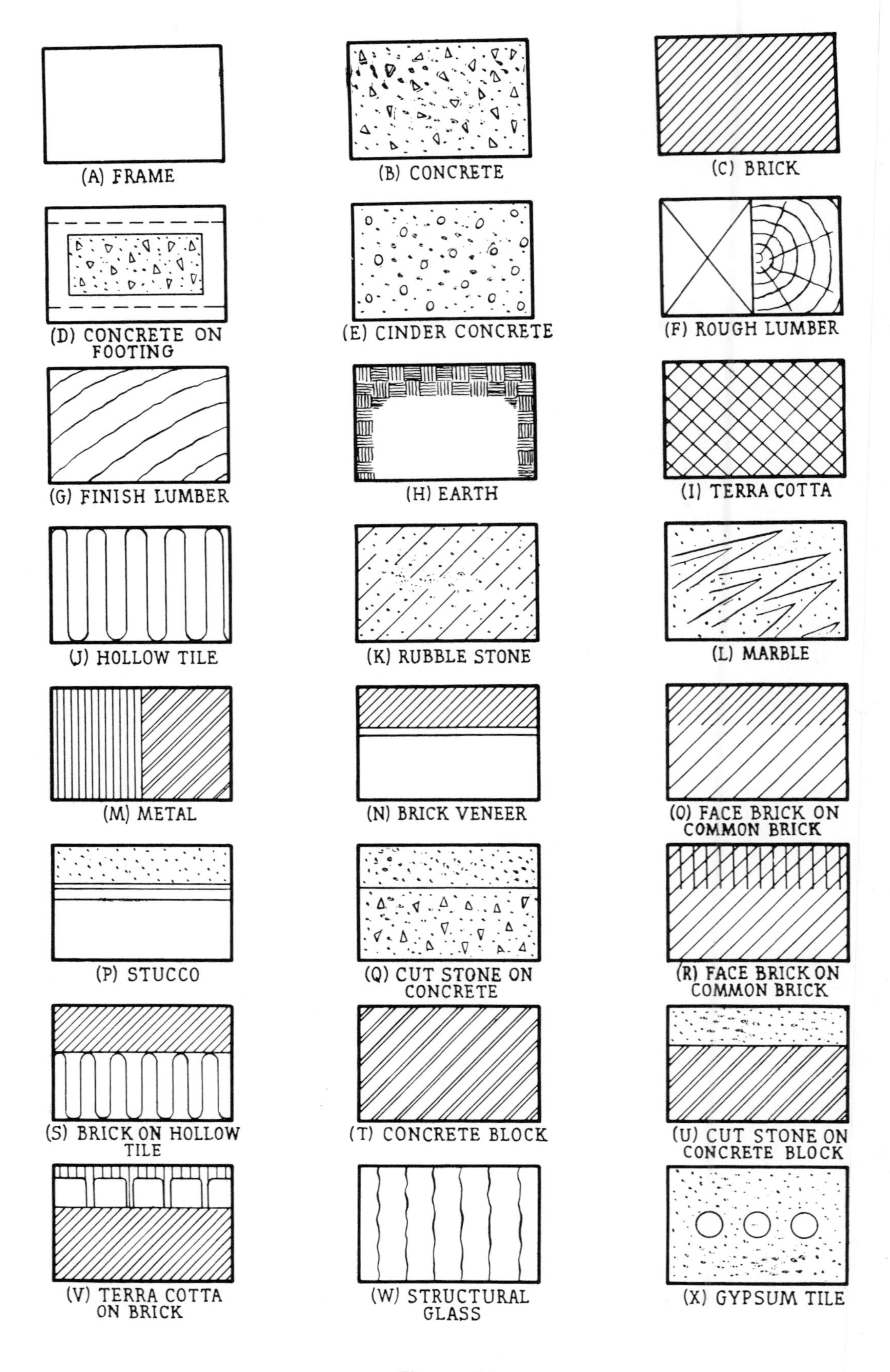

Figure 19

PLAN VIEW MATERIAL SYMBOLS

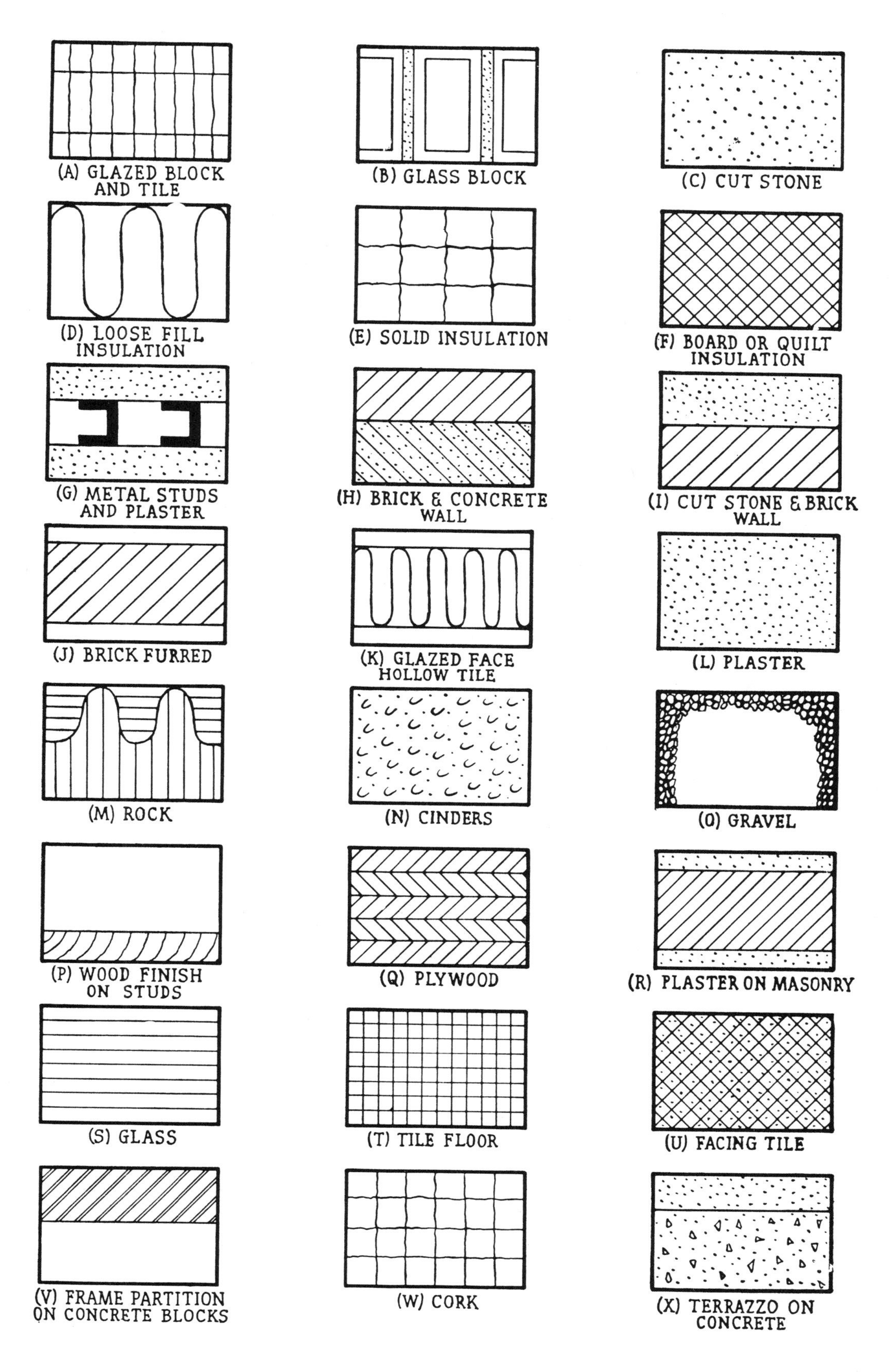

Figure 20

parallel lines. Many other symbols, as shown in Figures 19 and 20, can also be used for exterior walls.

Interior partitions, such as noted in the *F* part of Figure 16, can also be made of any one of many materials or combinations of several. If partitions are to be made of wood, the symbol shown at *A* in Figure 19 is used; if they are to be made of glass blocks, the symbol shown at *B* in Figure 20 is used.

Foundations Like walls and partitions, foundations are indicated by parallel lines whose distance apart, according to accurate scale, indicates required thickness. If the foundations are to be made of cast-in-place concrete, the symbol shown at *B* in Figure 19 would be between the parallel lines: if the foundations are to be made of concrete block, the symbol shown at *T* in the same illustration would be used.

Footings Unless otherwise noted in the prints, footings are always made of cast-in-place concrete. When such footings are to be used under cast-in-place concrete foundations, the symbol shown at *D* in Figure 19 is used to indicate both footings and foundations. The distance between the dashed lines indicates the width of the footings. The width of the concrete symbol indicates the thickness of the foundation.

Insulation In modern structures, a considerable amount of this material is used as a means of resisting heat loss during the winter and heat gain during the summer. The symbols are shown at *D*, *E*, *F*, and *W* in Figure 20.

Tile Floors Where tile floors are required, as in bathrooms, the symbol shown at *T* in Figure 20 is shown. The size and kind of tile is explained in specs.

Wood Floors When wood floors are required, no symbol is shown. In such cases, it is understood that the specs will describe the type and kind of wood to be used. If a special type of wood floor is required, further details may be shown.

The symbols shown in Figures 19 and 20 are self-explanatory. However, it should be kept in mind that they are drawn accurately to scale and that, if necessary, they can be scaled. Many of the symbols will also be used in connection with the subject of Chapter 5.

Uncommon Symbols When uncommon symbols are used in plan views, designers always include a symbol key to identify materials and details.

Detail Symbols As previously explained, all of the items shown in plan views must be indicated by symbols. Such symbols seldom look like the items they represent. The following explanations describe the commonly encountered detail symbols.

Fireplaces Following the idea of imagining that we can cut through items in order to see what they look like on the inside, let us also imagine that fireplaces can be cut and that we can look down at the cut surface. The *A* part of Figure 21 shows a typical fireplace symbol. All parts of the symbol are drawn to scale so that, if necessary, they can be measured. Note that many of the symbols shown in Figures 6, 19, and 20 are employed to make the complete fireplace symbol.

Chimneys The *B* part of Figure 21 shows a typical chimney symbol. This particular symbol is shown in connection with a brick wall. The number of flues may vary according to the needs of a particular structure.

Stairs Stairs are always indicated by the parallel lines shown at *w* and *y* in the *C* part of the illustration. The distance between these lines, according to scale, indicates the width of the stairs. Risers and treads are indicated as shown. The broken line, shown at *x*, indicates that one flight of stairs is above another. The "UP" stairs may go to a second floor and the "DN" stairs to a basement. The *D* part of the illustration shows the symbol necessary when closet space is to be immediately above stairs.

Areaways At *E*, a typical cast-in-place concrete areaway for a basement window is shown.

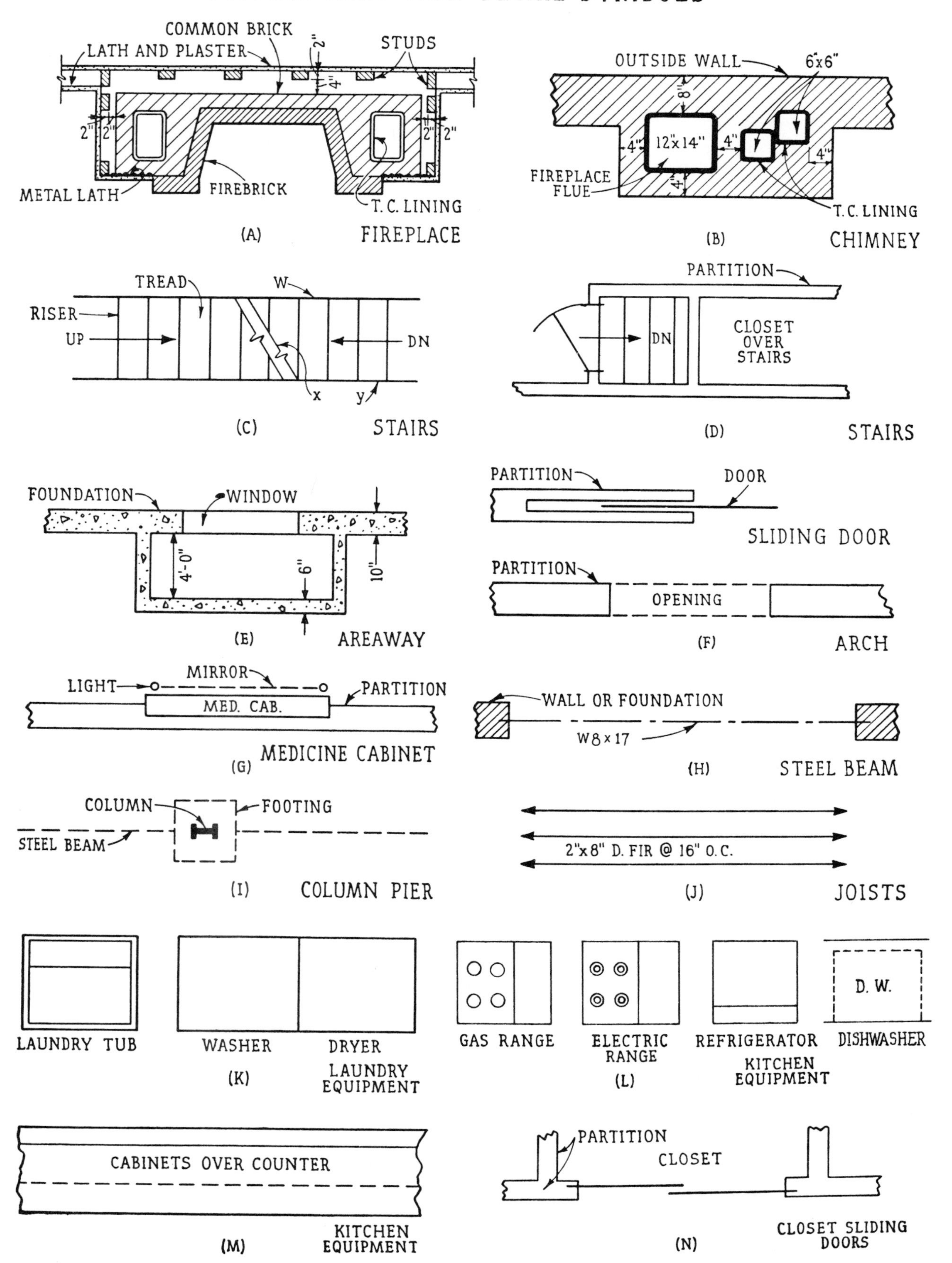
TYPICAL PLAN VIEW DETAIL SYMBOLS
COMMON BRICK
LATH AND PLASTER
STUDS
2"
4"
2"
2"
2"
2"
METAL LATH
FIREBRICK
T. C. LINING
(A)
FIREPLACE
OUTSIDE WALL
6"x6"
8"
12"x14"
4"
4"
4"
4"
FIREPLACE FLUE
T. C. LINING
(B)
CHIMNEY
TREAD
W
RISER
UP
DN
x
y
(C)
STAIRS
PARTITION
DN
CLOSET OVER STAIRS
(D)
STAIRS
FOUNDATION
WINDOW
4'-0"
6"
10"
(E)
AREAWAY
PARTITION
DOOR
SLIDING DOOR
PARTITION
OPENING
(F)
ARCH
MIRROR
LIGHT
PARTITION
MED. CAB.
(G)
MEDICINE CABINET
WALL OR FOUNDATION
W8 x 17
(H)
STEEL BEAM
COLUMN
FOOTING
STEEL BEAM
(I)
COLUMN PIER
2"x8" D. FIR @ 16" O.C.
(J)
JOISTS
LAUNDRY TUB
WASHER
DRYER
(K)
LAUNDRY EQUIPMENT
GAS RANGE
ELECTRIC RANGE
REFRIGERATOR
D. W.
DISHWASHER
(L)
KITCHEN EQUIPMENT
CABINETS OVER COUNTER
(M)
KITCHEN EQUIPMENT
PARTITION
CLOSET
(N)
CLOSET SLIDING DOORS

Figure 21

Areaways are used when basement windows are below grade.

Sliding Doors and Arches At *F*, the symbols for sliding doors and plastered arches are shown. Note that frame walls are indicated. A plastered arch is an opening, something like a door, in a wall. The arch generally has a curved top.

Medicine Cabinet The symbol at *G* shows a typical way of indicating a medicine cabinet in a frame partition.

Steel Beams The broken line at *H* represents a steel beam which is used in basements to support floors and partitions above. The notation means that the beam is 8 inches deep and weighs 17 pounds per lineal foot.

Joists The series of double arrows at *J* shows the symbol used to specify size, kind, and spacing of floor joists. The notation means that the joists are to be 2 inches thick and 8 inches deep, made of Douglas fir, and spaced 16 inches on center.

Dimensioning Walls and Partitions The top part of Figure 22 shows typical ways in which dimension lines are drawn for several types of exterior and interior walls and partitions. Various designers use somewhat different methods of dimensioning but the practice shown is typical enough for our purpose.

Doors and Windows Symbols in Figure 22 also show how doors and windows, of various kinds, are set forth and dimensioned in brick and frame exterior walls. Here we can see the great difference between door and window symbols in elevation and plan views. As previously explained, in connection with Figure 16, plan-view symbols appear as we would see them if we could cut an existing house in half, move the top part away, and look straight down at the cut surfaces of the bottom part.

Doors and windows in masonry walls are dimensioned to their centers because their frames are set up first and the masonry built around them. The swing of doors and casement windows is indicated by a small arc and a line, opening out or in as marked. The D.H. (double-hung) window slides up and down. Door sizes are shown in two ways. In some cases the doors are numbered or lettered in a circle which refers to a door schedule. In other cases, the door sizes are put on or near the line indicating the swing of the door. Window sizes are also shown in two ways. As for doors, they are sometimes numbered or lettered in a circle which refers to a schedule. In other cases, the sizes are shown near the symbols, in either plan or elevation views.

Also shown in the illustration are symbols for alterations (remodeling), the pattern for a brick walk laid on a concrete slab, a typical finish schedule which tells the materials for the finished work in each room of a house, and some sketches showing how a brick wall is built (bonded) as a means of giving it strength.

Figure 23. This illustration shows some typical plumbing, heating, ventilating, and electrical symbols. All mechanics should be familiar with these symbols, even though they do not install the equipment, so that sufficient space can be allowed and provisions made for the proper connections. The plumbing, heating, and ventilating symbols are drawn accurately to scale so that the space required can be ascertained while drawings are being made and so that mechanics will know exactly how much space to allow for the equipment.

Electrical outlets are indicated on plan views in approximate locations. For fixtures, the capital letter in the symbol indicates the fixture type, the lower-case letter designates the switch control, and the number identifies the electrical circuit. In the switch symbol, the letter indicates the controlled device. Architects use a variety of electrical symbols. However, once we learn to recognize and understand the symbols shown in the illustration, we should also understand any variations.

NOTES

In many instances, drawings contain printed instructions relative to any one or more of

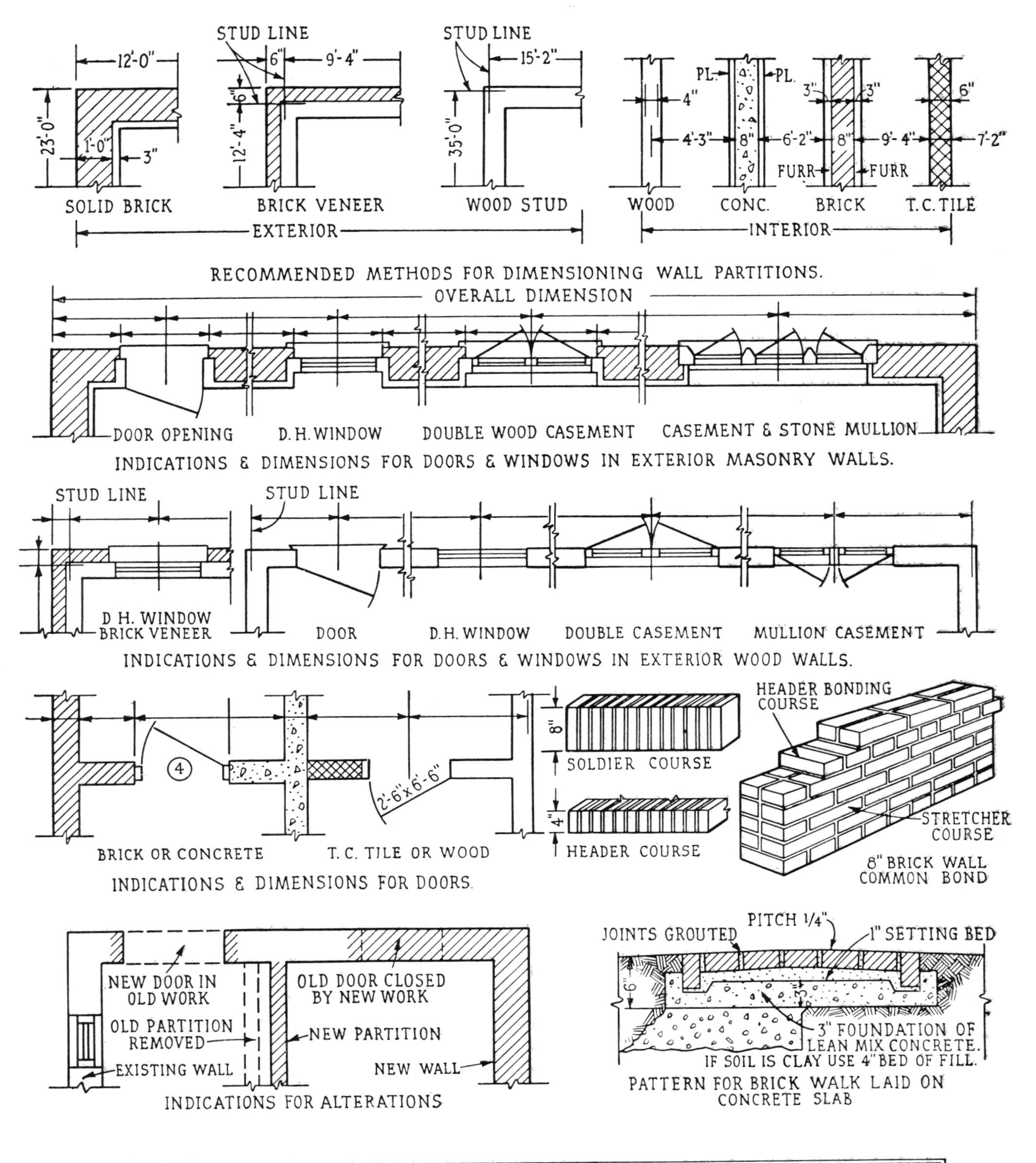

FINISH SCHEDULE

NUMBER	ROOM	WALLS	CEILING	DADO	BASE	FLOOR	CORNICE	NOTES
101	LIV. ROOM	PLASTER	PLASTER	WOOD	WOOD	OAK	WOOD	BOOKCASE
102	LAVATORY	"	"	TILE	TILE	TILE	———	———
103	HALL	"	"	———	WOOD	OAK	PICT. MOLD	SEE DETAIL
104	KITCHEN	"	"	———	RUBBER	LINOLEUM	———	———

Figure 22

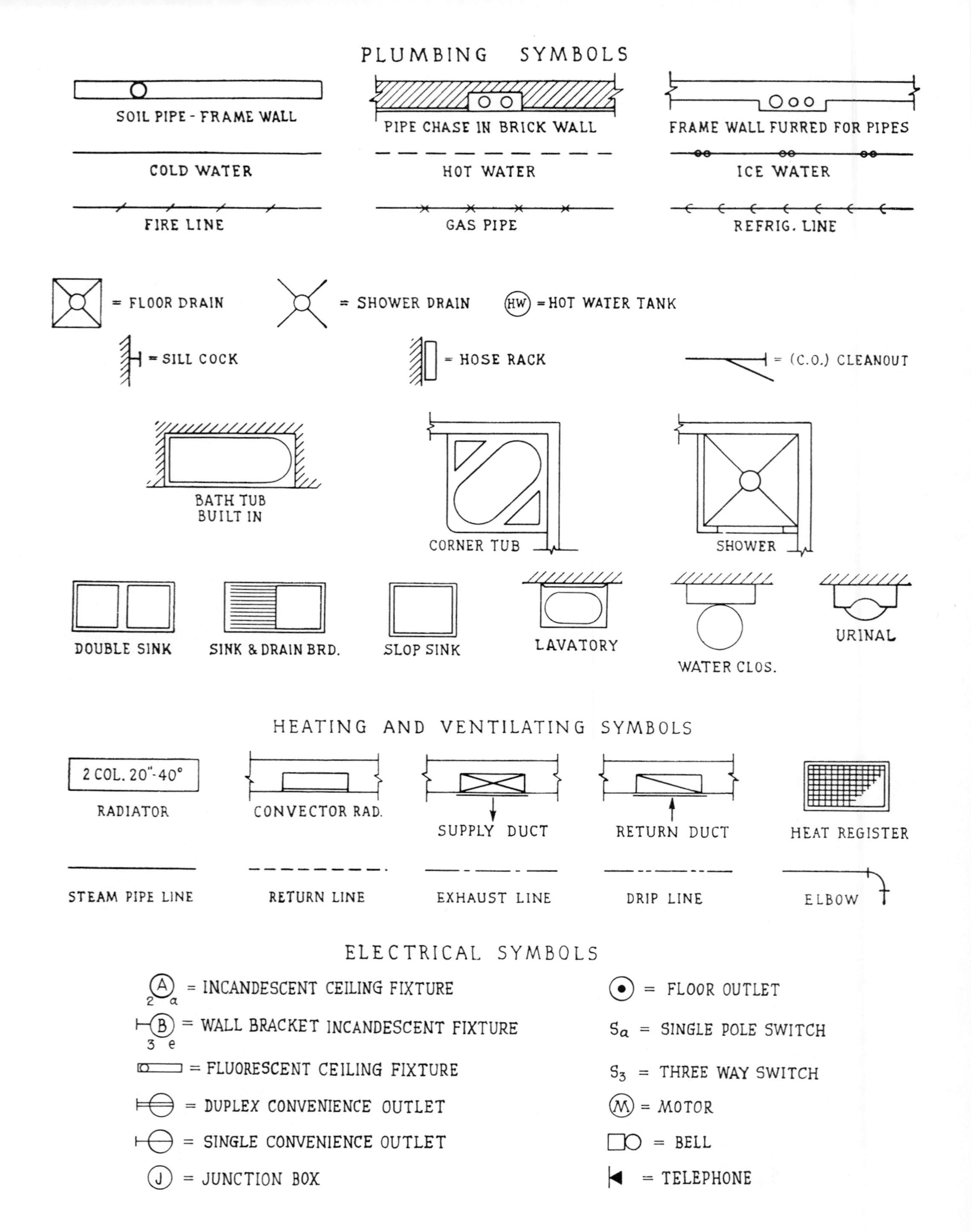
PLUMBING SYMBOLS
SOIL PIPE - FRAME WALL
PIPE CHASE IN BRICK WALL
FRAME WALL FURRED FOR PIPES
COLD WATER
HOT WATER
ICE WATER
FIRE LINE
GAS PIPE
REFRIG. LINE
= FLOOR DRAIN
= SHOWER DRAIN
HW
= HOT WATER TANK
= SILL COCK
= HOSE RACK
= (C.O.) CLEANOUT
BATH TUB
BUILT IN
CORNER TUB
SHOWER
DOUBLE SINK
SINK & DRAIN BRD.
SLOP SINK
LAVATORY
WATER CLOS.
URINAL
HEATING AND VENTILATING SYMBOLS
2 COL. 20"-40°
RADIATOR
CONVECTOR RAD.
SUPPLY DUCT
RETURN DUCT
HEAT REGISTER
STEAM PIPE LINE
RETURN LINE
EXHAUST LINE
DRIP LINE
ELBOW
ELECTRICAL SYMBOLS
A
2 a
= INCANDESCENT CEILING FIXTURE
B
3 e
= WALL BRACKET INCANDESCENT FIXTURE
= FLUORESCENT CEILING FIXTURE
= DUPLEX CONVENIENCE OUTLET
= SINGLE CONVENIENCE OUTLET
J
= JUNCTION BOX
= FLOOR OUTLET
Sa = SINGLE POLE SWITCH
S3 = THREE WAY SWITCH
M
= MOTOR
= BELL
= TELEPHONE

Figure 23

several details of construction. Such notes either give additional information or refer to other parts of the construction plans where more information can be found. Therefore, we should pay careful attention to all notes on drawings.

LINES

Designers generally use a variety of special types of lines for particular purposes on drawings or parts of drawings. If we learn to recognize such lines, we can more quickly and better visualize drawings.

Broken Lines This line has a wavy break in it at intervals and is used to indicate that parts of a drawing have been left out or that the full length of some part has not been shown.

Invisible Lines This kind of line is one made up of a series of short dashes and is used to indicate hidden or invisible edges—edges that are hidden under some other part of a structure. The dashed lines shown in the *E* part of Figure 4 constitute a good example of this kind of line and its purpose.

Center Lines Center lines are made up of alternating long and short dashes and are generally called *dash-and-dot* lines. The line is drawn light and used to indicate or locate centers.

Section Lines A section line, sometimes known as a *reference line*, is solid and has an arrowhead at each end pointing in the direction of a certain part of a house which is shown in greater detail on another sheet of the plans. Letters or numbers are shown at the arrow ends of the lines. We shall become better acquainted with this type of line when we study Chapter 5.

TERMS

ACCESS PANEL: A removable wall panel which allows access to plumbing and other equipment.

AGGREGATE: The stone and sand used in making concrete.

ALUMINUM INSULATION: Aluminum foil shaped to fit between joists and studs (see *G* part of Figure 25).

ANCHOR BOLTS: Bolts used to anchor the sill (or mud sill) of a structure to the foundation (see *B* part of Figure 24).

ARCH: A wall opening, without door, whose top is arched or elliptical in shape.

AREAWAY: An open area outside of basement windows, used when windows are below grade level (see *G* part of Figure 24).

ASH DUMP: An opening in a fireplace hearth through which ashes may be disposed of (see *D* part of Figure 25).

BACK HEARTH: The part of a hearth which is inside a fireplace (see *C* part of Figure 24).

BALUSTER: The spindles or vertical parts of a railing (see *I* part of Figure 24).

BASEBOARD: A molding placed on walls near the floor.

BEAM: A horizontal steel or wood member used to support floors and other structural parts (see *H* part of Figure 24 and *E* part of Figure 25).

BLANKET INSULATION: A wool-type insulation which is made in blanketlike form ready for installation between joists and studs (see *G* part of Figure 25).

BOARD INSULATION: Rigid insulation made into large boardlike panels (see *H* part of Figure 25).

BRIDGING: Wood or steel brackets used to keep floor joists in vertical position (see *A* part of Figure 24).

BUILDING PAPER: A heavy waterproof paper used on the walls and in other parts of a structure.

CAMBER: The convex curve of the edge of a joist.

CATCH BASIN: A receptacle into which water from roofs and areaways drains.

COLLAR BEAM: A 2 by 4 or 2 by 6 used to stiffen a roof (see *I* part of Figure 25).

CORNER POST: Two or more studs spiked together to form a corner between wall frames (see *B* part of Figure 24).

DAMPPROOFING: A waterproof material used to prevent moisture from getting through foundations and other structural members (see *D* part of Figure 24).

DRAIN: A means of carrying off water (see *G* part of Figure 24).

DUCT: Pipes, usually made of sheet steel, round or rectangular, and used in heating and air conditioning systems.

FIELD TILE: Tile without bell connections used to drain off water from the bottoms of foundations and footings.

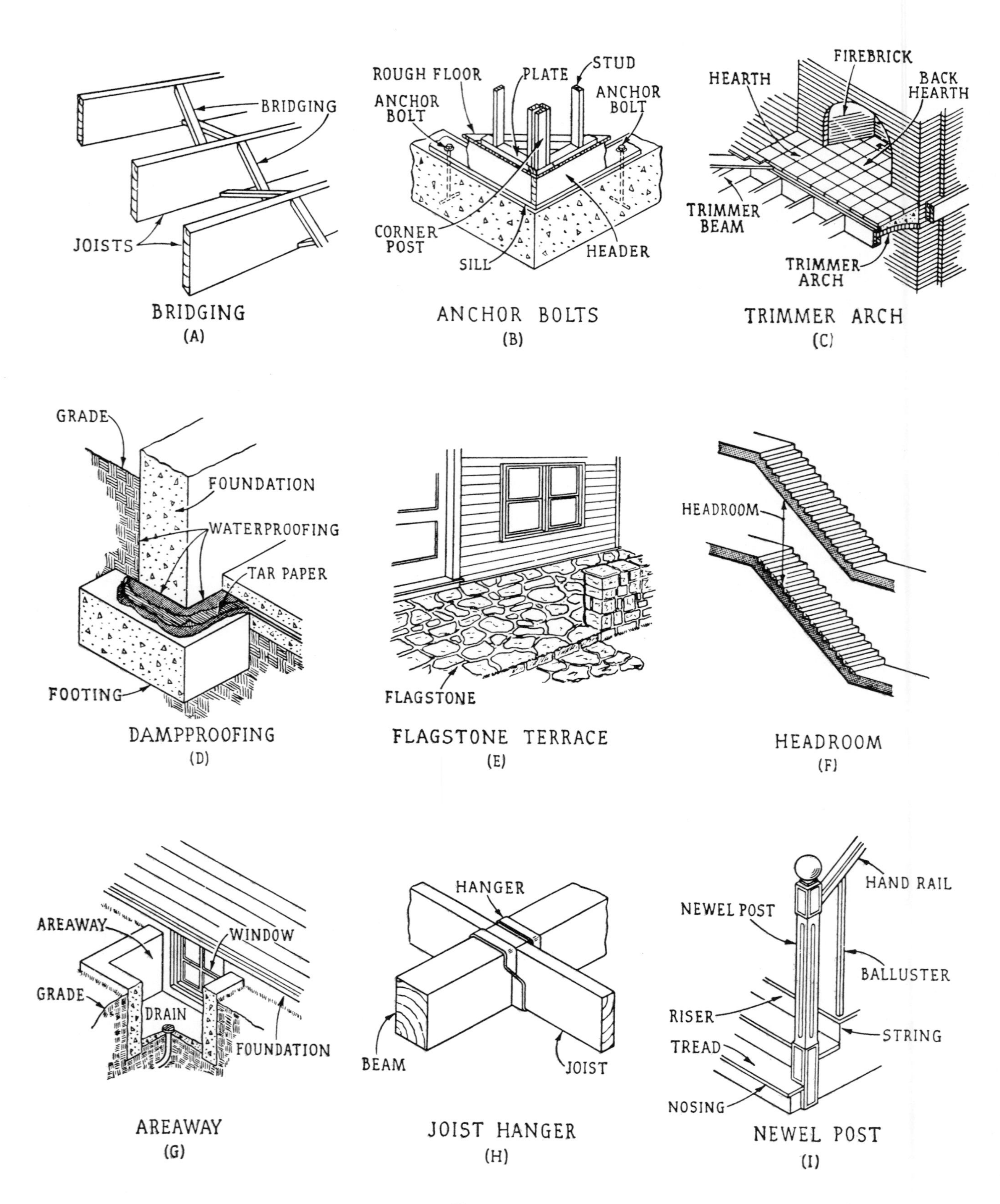
BRIDGING
JOISTS
BRIDGING
(A)
ROUGH FLOOR
PLATE
STUD
ANCHOR BOLT
ANCHOR BOLT
CORNER POST
SILL
HEADER
ANCHOR BOLTS
(B)
FIREBRICK
HEARTH
BACK HEARTH
TRIMMER BEAM
TRIMMER ARCH
TRIMMER ARCH
(C)
GRADE
FOUNDATION
WATERPROOFING
TAR PAPER
FOOTING
DAMPPROOFING
(D)
FLAGSTONE
FLAGSTONE TERRACE
(E)
HEADROOM
HEADROOM
(F)
AREAWAY
WINDOW
GRADE
DRAIN
FOUNDATION
AREAWAY
(G)
HANGER
BEAM
JOIST
JOIST HANGER
(H)
HAND RAIL
NEWEL POST
BALLUSTER
RISER
STRING
TREAD
NOSING
NEWEL POST
(I)

Figure 24

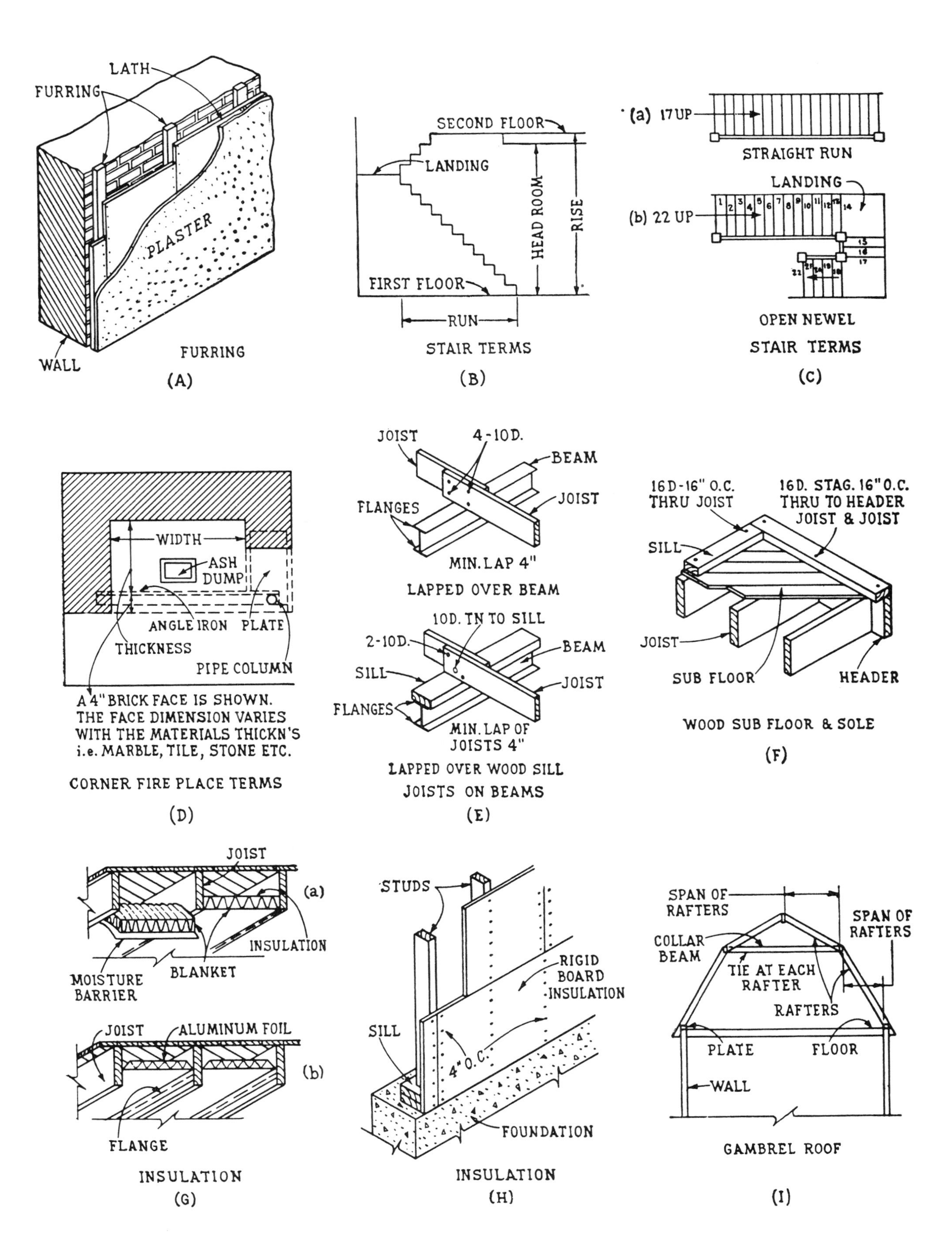

LATH
FURRING
PLASTER
WALL
FURRING
(A)
SECOND FLOOR
LANDING
HEAD ROOM
RISE
FIRST FLOOR
RUN
STAIR TERMS
(B)
(a) 17 UP
STRAIGHT RUN
LANDING
(b) 22 UP
OPEN NEWEL
STAIR TERMS
(C)
WIDTH
ASH DUMP
ANGLE IRON
PLATE
THICKNESS
PIPE COLUMN
A 4" BRICK FACE IS SHOWN.
THE FACE DIMENSION VARIES
WITH THE MATERIALS THICKN'S
i.e. MARBLE, TILE, STONE ETC.
CORNER FIRE PLACE TERMS
(D)
JOIST
4-10D.
BEAM
JOIST
FLANGES
MIN. LAP 4"
LAPPED OVER BEAM
10D. TN TO SILL
2-10D.
BEAM
SILL
JOIST
FLANGES
MIN. LAP OF JOISTS 4"
LAPPED OVER WOOD SILL
JOISTS ON BEAMS
(E)
16D-16" O.C. THRU JOIST
16D. STAG. 16" O.C. THRU TO HEADER JOIST & JOIST
SILL
JOIST
SUB FLOOR
HEADER
WOOD SUB FLOOR & SOLE
(F)
JOIST
(a)
INSULATION
MOISTURE BARRIER
BLANKET
JOIST
ALUMINUM FOIL
(b)
FLANGE
INSULATION
(G)
STUDS
RIGID BOARD INSULATION
SILL
4" O.C.
FOUNDATION
INSULATION
(H)
SPAN OF RAFTERS
SPAN OF RAFTERS
COLLAR BEAM
TIE AT EACH RAFTER
RAFTERS
PLATE
FLOOR
WALL
GAMBREL ROOF
(I)

Figure 25

FIREBRICK: A special type of brick which can stand great heat. Used in fireplaces (see *C* part of Figure 24).

FLAGSTONES: Flat stones of irregular shape used as flooring for walks and exterior living spaces (see *E* part of Figure 24).

FLANGE: The upper and lower cross parts of a steel I-beam (see *E* part of Figure 25).

FLOOR SLAB: A floor made of concrete.

FLUE: Passageway in chimneys for carrying off smoke and gases.

FOOTING: The spread portion at the bottom of a foundation of column to prevent settling (see *D* part of Figure 24).

FOUNDATION: The masonry walls which support the framework of a structure (see *G* part of Figure 24 and *H* part of Figure 25).

FRAMING: Rough carpentry work which forms the skeleton of a structure (see *B* part of Figure 24 and *F* and *H* parts of Figure 25).

FURRING: Strips of wood used to create an air space between plaster and wall surfaces. Sometimes used in floors (see *A* part of Figure 25).

GIRDER: A beam that supports other beams.

GIRT: The heavy horizontal timber carrying the joists in a frame house.

GRADE: The ground level around a structure (see *D* and *G* parts of Figure 24).

GROUNDS: Strips of wood around windows and other openings against which plaster is butted and to which trim is nailed.

HANDRAIL: Portion of a stair railing (see *I* part of Figure 24).

HEADER: A wood member used at the ends of joists (see *B* part of Figure 24 and *F* part of Figure 25).

HEADROOM: The distance between flights of steps or between steps and ceiling above (see *F* part of Figure 24 and *B* part of Figure 25).

HEARTH: The area in front of a fireplace (see *C* part of Figure 24).

HOSE BIB: An exterior faucet for connecting a hose.

HOUSE DRAIN: The piping that carries off the discharge from all soil and waste lines in a house.

HOUSE SEWER: The drainage pipe connecting with the drain about 10 feet outside the house.

INSULATION: A special type of building material used to resist heat loss and heat gain through walls and other structural parts (see *G* and *H* parts of Figure 25).

JOIST HANGER: A metal device used to support the ends of joists (see *H* part of Figure 24).

JOISTS: Lightweight beams used to support floors (see *E*, *F*, and *G* parts of Figure 25).

KNEE WALL: An interior partition which is less than room height. Used between floors and sloping roofs.

LALLY COLUMN: A cylindrically shaped steel member, sometimes filled with concrete, used as a support for girders or other beams.

MOISTURE BARRIER: A special coating applied to insulation to make it moistureproof (see *G* part of Figure 25).

NEWEL: An upright post supporting the handrail at the top and bottom of a stairway, or at the turns on a landing; also the main post about which a circular staircase winds, sometimes called a newel post (see *I* part of Figure 24).

OPEN NEWEL STAIRS: A staircase which runs in three directions (see *C* part of Figure 25).

PIER: A rectangular or square masonry support or footing either built into a wall or free standing.

PIPE COLUMN: A pipelike column such as is used to support corners of corner fireplaces (see *D* part of Figure 25).

PLATE: A horizontal member in wall framing.

RAFTER: The sloping or almost horizontal members which support pitched and flat roofs (see *I* part of Figure 25).

RAFTER SPAN: The horizontal distance a rafter covers (see *I* part of Figure 25).

REINFORCED CONCRETE: Concrete which is made stronger by the use of steel.

SCUTTLE: An opening in a ceiling which provides access to areas under roofs.

SILLS: Bottom members of windows and doors; also used to support lower ends of wall studs (see *B* part of Figure 24 and *F* and *H* parts of Figure 25).

SOIL STACK: Pipe which receives discharge from all plumbing fixtures.

STAIR LANDING: A horizontal floor area between flights of steps (see *C* part of Figure 25).

STAIR RISE: The vertical distance between the top and bottom of a staircase (see *B* part of Figure 25).

STAIR RUN: The horizontal distance required for a staircase (see *B* part of Figure 25).

STRAIGHT-RUN STAIRS: A staircase without a twist or turn (see *C* part of Figure 25).

STUDS: Small posts, usually 2 by 4 or 2 by 6, used as vertical framing in walls and partitions (see *H* part of Figure 25).

SUBFLOOR: A rough floor applied directly to joists (see *F* part of Figure 25).

TERRACE: The floor of a patio or other exterior area (see *E* part of Figure 24).

TRIMMER ARCH: A masonry arch used to support the hearth in a fireplace (see *C* part of Figure 24).

TRIMMER BEAM: Two joists spiked together. Used around openings in floors (see *C* part of Figure 24).

TYPES OF HOUSES

The plan views to be illustrated and explained in the following section of this chapter are directly related to the elevation views which were illustrated and explained in a comparable section of Chapter 3. For example, the plan views shown in Figures 26*A* and 26*B* of this chapter, and the elevation views shown in Figure 11 of Chapter 3, all represent the same structure.

Visualizing the relationship between the plan and elevation views, shown in this chapter and

ABBREVIATIONS

The following abbreviations constitute the most commonly encountered ones. There are hundreds more. Readers can ask about any uncommon abbreviations they may find on drawings. Abbreviations are also used, to a great extent, to suit the particular fancy of the person preparing the drawings. Thus, there is no way of standardizing them or of showing any and all possible variations.

Air Conditioner	A.C.
Apartment	Apt.
Bathroom	B.
Beam	Bm.
Bearing Plate	Brg. Pl.
Bedroom	B.R.
Block	Bl.
Cast Stone	CS
Catch Basin	C.B.
Ceiling	Ceil.
Center to Center	C. to C.
Cleanout	C.O.
Closet	Clos., Cl., C.
Clothes Chute	C.C.
Cold Air	C.A.
Cold Water	C.W.
Collar Beam	Col. B.
Column	Col.
Conduit	Cond.
Cubic Feet	Cu. Ft.
Detail	Det.
Dining Room	D.R.
Dishwasher	D.W.
Double-acting Door	D.A.D.
Double-strength Glass	D.S.A.
Drawings	Drgs.
Dressed and Matched	D and M
Finish	Fin.
Firebrick	F.B.
Floor Drain	F.D.
Flush	Fl.
Frame	Frm.
Folding Door	F.D.
Folding Partition	F.P.
Furring	Fur.
Garage	Gar., G.
Gas	Gas
Gas Range	G.R.
Hall	H.
Heavy Weight	H.W.
Hose Bib	H.B.
Hot Air	H.A.
Hot Water	H.W.
House Drain	H.D.
Jamb	Jb.
Joist Space	J.S.
Kitchen	K.
Kitchen Cabinet	K.C.
Knocked Down	K.D.
Lally Column	Lal. Col.
Laundry	L.
Lavatory	Lav.
Lineal	Lin.
Lining	Lin.
Linoleum	Linol.
Living Room	L.R.
Marble Threshold	M.T.
Masonry Opening	M.O.
Maximum	Max.
Medicine Cabinet	M.C.
Mixture	Mix.
Mortar	Mor.
Nosing	Nos.
On Centers	O.C.
Panel	Pan.
Patio	Pat.
Penny (nails)	d
Picture Molding	P.M.
Plastered Opening	P.O.
Plate	Pl.
Plumbing	P.
Pull Switch	P.S.
Radiator	R
Recessed	Rec.
Refrigerator	R.
Riser	R.
Room	Rm.
Screen	Scr.
Section	Sec.
Shelving	Shelv.
Single-strength Glass	S.S.A.
Sliding Door	Sl. Dr.
Square Feet	Sq. Ft.
Standard Door	S.D.
Standard Weight	S.W.
Steel Sash	S.S.
Threshold	Th.
Tread	T.
Two-member	2-M
Utility Room	U.R.
Water Closet	W.C.
Wide-flange Beam	W.
Window Radiator	W.R.
Wrought Iron	W.I.
Yellow Pine	Y.P.

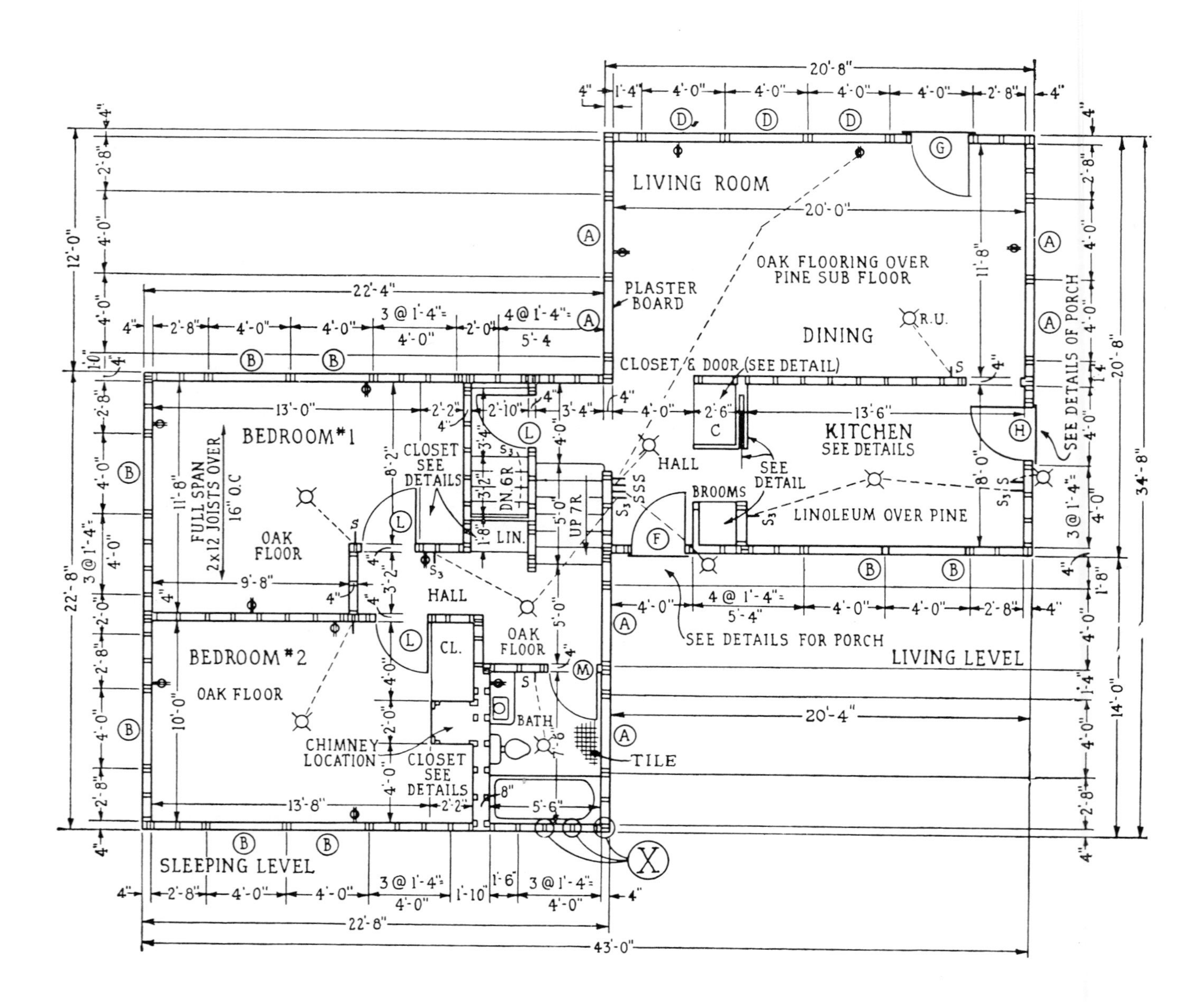

DOOR SCHEDULE		
MARK	SIZE	REMARKS
F	1¾"x 3'-0"x 7'-0"	FLUSH PANEL
G	1¾"x 2'-8"x 7'-0"	FLUSH PANEL & SIDE LIGHT (SEE DTL.)
H	1¾"x 2'-8"x 7'-0"	GLASS PANEL 24"x28"
J	1¾"x 3'-0"x 6'-8"	" " "
K	1¾"x 3'-0"x 6'-8"	SOLID
L	1⅜"x 2'-6"x 6'-8"	FLUSH PANEL (BEDRM.)
M	1⅜"x 2'-4"x 6'-8"	" " (BATH)
N	1⅜"x 2'-6"x 6'-8"	SOLID
P	8'-0" x 6'-6"	GARAGE DOOR

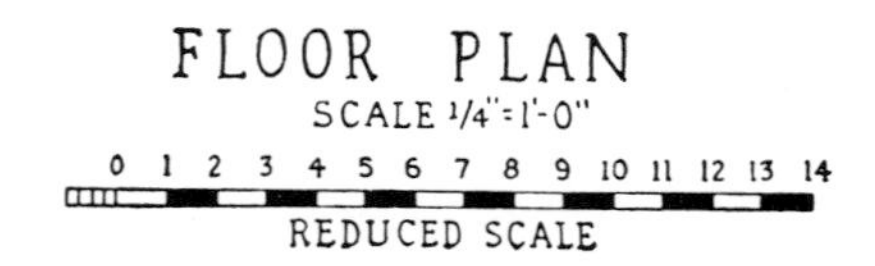

Figure 26A

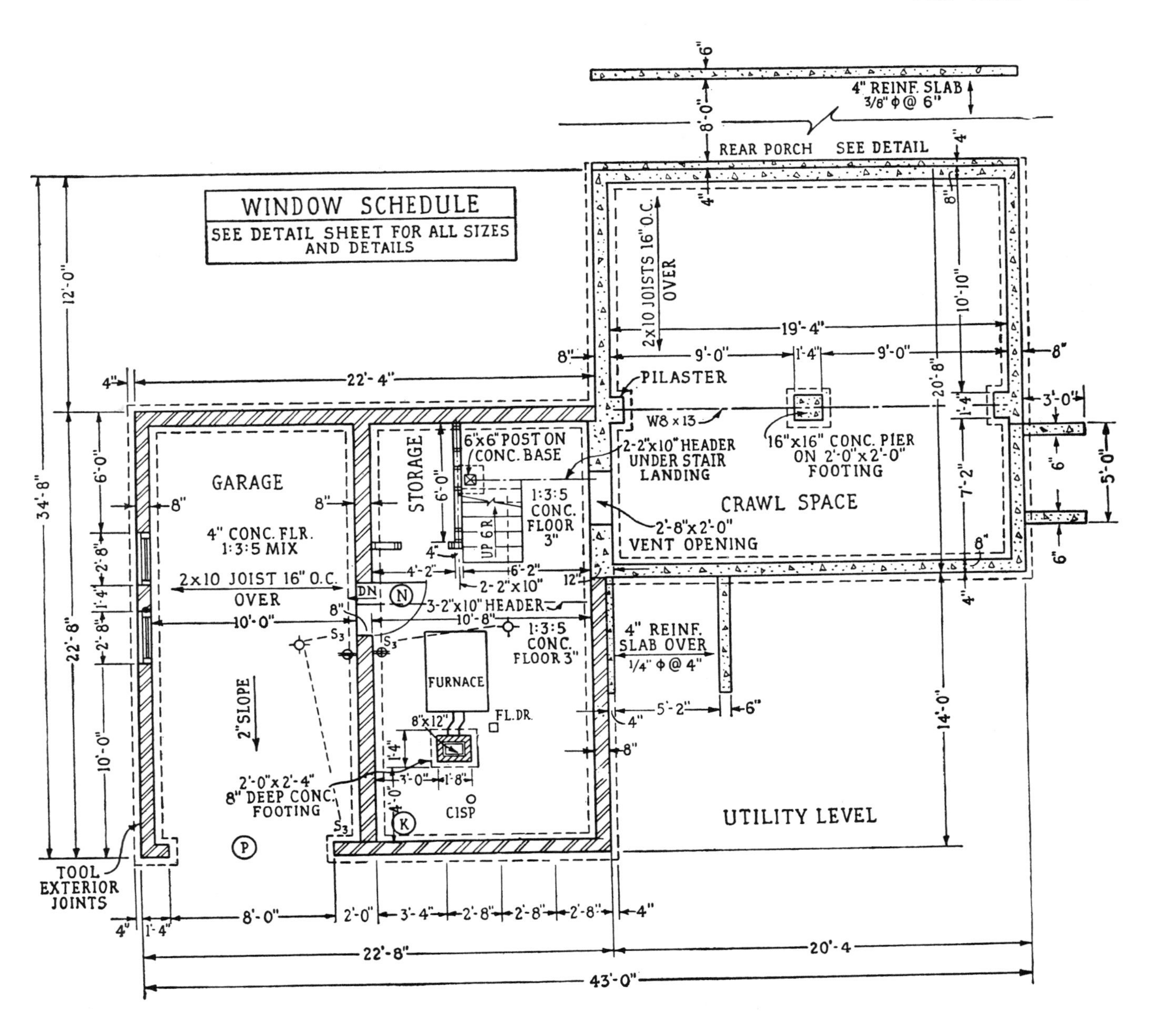

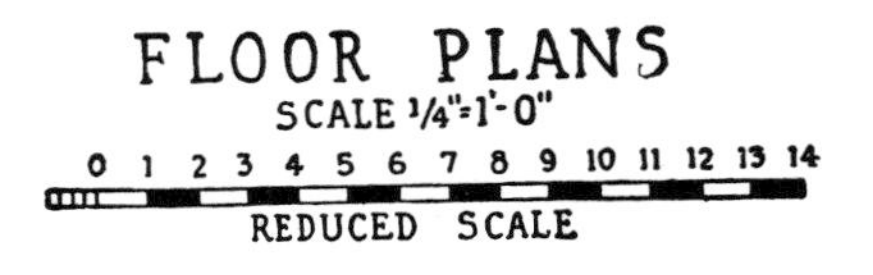

Figure 26B

in Chapter 3, can be accomplished with ease if the previously explained process is followed. For example, imagine that the plan views can be placed *above* the elevation views and that projection lines can be drawn as explained in relation to Figure 18. We cannot very well tear pages out of this book in order to place the plan views above the elevation views. However, if need be to aid our visualization, we can use tissue paper and a pencil to make rough tracings of the plan views. The tracings can be placed above the elevation views so that the relationship between the two views can be visualized and understood.

There are other ways by which we can help ourselves to visualize relationships between plan and elevation views. For example, select an exterior door on a plan view. Note the letter or number used to identify it. Then find the door having the same letter or number on one of the elevation views. In like manner, select windows, porches, etc. on plan views and find these same items on the elevation views.

The plan view shown in Figure 26*A* contains a symbol which is seldom used in connection with plan views. Note the circle marked *X* which is shown with the exterior wall symbol for the bathroom. The sets of parallel lines, perpendicular to the lines of the wall symbol, represent the studs. In this case, the architect has planned where each and every stud is to be placed.

There are several reasons why stud locations are sometimes planned by the designer instead of being left to the judgment of the builders. Sometimes a particular type wall sheathing or interior finish requires special stud locations. In other cases, preparation of framing members, such as studs, is carried on before actual construction gets under way. Or the window locations may be specified on detail drawings in such a manner that exact stud locations are required.

Another uncommon practice is also shown in connection with Figure 26*B*. The window schedule refers to a detail sheet where all necessary information is given.

A final example of the use of uncommon symbols has to do with the chimney location shown in the bedroom 2 area, also in Figure 26*A*. The designer has omitted the usual chimney symbol and has shown only the stud locations and wall surfaces around the area required by the chimney.

The plan and elevation views various designers prepare may include uncommon symbols or methods of indicating necessary information. However, we can avoid any chance of confusion or trouble by scanning any drawings we encounter with the idea of learning, first of all, the general presentation plan being employed.

CONSTRUCTION-PLAN STUDIES

Figures 26*A*, 26*B*, 27*A*, 27*B*, and 28 represent actual plan views as architects prepare them and as builders use them. In this book, because of page-size limitations, the plan views have to be shown on a reduced scale. We can study the drawings just as though they were regular size to learn what they intend to have built.

Figures 26*A* and 26*B*. The following observations will be helpful in learning to read the plan views for a split-level house. Readers are urged to locate all the symbols, as they are mentioned, and to make sure that every explanation is thoroughly understood. This constitutes a most excellent practice of the kind necessary when learning to read plans.

Materials The plan-view symbols show what various structural items are to be made of:

Chimney	*brick*	Beam pier	*concrete*
Footings	*concrete*	Partitions	*frame*
Pilasters	*concrete*	Reinforcing	*steel*
Column posts	*wood*	B. R. floor	*oak*
L. R. floor	*oak*	Bath floor	*tile*
Interior finish	*gypsum-board*	Kitchen floor	*linoleum*
		Garage floor	*concrete*
Subfloor	*pine*	Crawl-space foundation	*concrete*
Porch slab	*concrete*	Garage walls	*concrete block*
Porch foundations	*concrete*		

Details Several of the symbols, such as the ones associated with the porches, closets, and the kitchen, have notes near them which refer to detail sheets. We shall study details in a later chapter.

Windows The window schedule contains a

note to the effect that all information about windows is shown on a separate detail sheet.

Doors The door symbols, in both the elevation (see Figure 11) and plan views, contain circled letters which refer to a door schedule. Such a schedule is shown in the plan views of the sleeping and living levels. For example, door *F* must be 1¾ inches thick, 3′ 0″ wide, 7′ 0″ high, and of flush design. The door for bedroom 2 must be 1⅜ inches thick, 2′ 6″ wide, 6′ 8″ high, and of flush-panel design.

Levels Three levels are indicated and coincide with the levels shown on the elevation views in Figure 11.

Garage Walls All the walls surrounding the garage and heater areas must be made of concrete block. This information is indicated on both the plan and elevation views. A note indicates that the joints between blocks must be tooled. This means that the masons must make attractive joints so that the exterior surfaces of the wall will look better than ordinary foundations.

Crawl-space Foundations The foundations around the crawl space must be cast-in-place concrete. The pilasters and pier must also be made of cast-in-place concrete. Note that part of the foundation, between the crawl space and the heater area, is to be 12 inches thick. Note, too, that in the 12-inch part of the foundation, a 2′ 8″ by 2′ 0″ opening must be provided.

Stairs Between the living- and sleeping-area levels, a staircase containing 7 risers must be built. The span must be 5′ 0″ and the width 3′ 4″. Between the living- and utility-area levels, a staircase having 6 risers must be built. The span at the living-area level must be 3′ 2″ and the width must be 2′ 10″.

General In the living area there is to be a partition between the kitchen and the dining space. That partition is supported by an 8-inch steel beam which weighs 13 pounds per lineal foot. The midpoint of the beam is to be supported by a 16- by 16-inch concrete pier on a 2′ 0″ by 2′ 0″ footing. The ends of the beam must be supported by pilasters which are to be placed as part of the foundation. Pilasters are thickened portions of the foundation wall. They are needed to spread the load transmitted to the foundations by the ends of the steel beam. The wall footings are correspondingly increased in size around the pilasters.

The rear porch is to have a cast-in-place concrete slab floor 4 inches thick. The concrete must be reinforced by means of ⅜ inch round steel bars spaced 6 inches apart. The outer side of the slab is to be supported by a 6-inch cast-in-place concrete foundation which is to extend the full length of the porch. This information is shown in Figure 26*B*. The elevation view shows that the surface of the porch floor is to be 5 inches below the surface of the kitchen floor. The inner side of the slab is to be supported by a 4-inch-thick extension of the main foundation.

The porch in connection with door *F* is to have a reinforced cast-in-place concrete slab floor. This information is shown on the plan view of the utility level. The concrete is to be 4 inches thick and must be reinforced by means of ¼-inch round steel bars placed 4 inches apart. The outer side of the porch is to be supported by a 6-inch cast-in-place concrete foundation, as indicated by the plan view of the utility level. The inner side is to be supported by a 4-inch cast-in-place concrete extension of one of the heater-room walls.

The front elevation view indicates that a small cast-in-place concrete porch is required in connection with door *H*. The information concerning that porch and other such porches can be read in like manner.

Dimensions The plan views show dimensions which indicate the required positions of many exterior wall studs. For example, note the rear-wall symbol for bedroom 1. Starting at the left-hand corner, a 2′ 8″ dimension locates two studs. Between those studs and the corner post (to be made of three studs) one other stud is indicated. Two 4′ 0″ dimensions are used to locate the center line between the two *B*

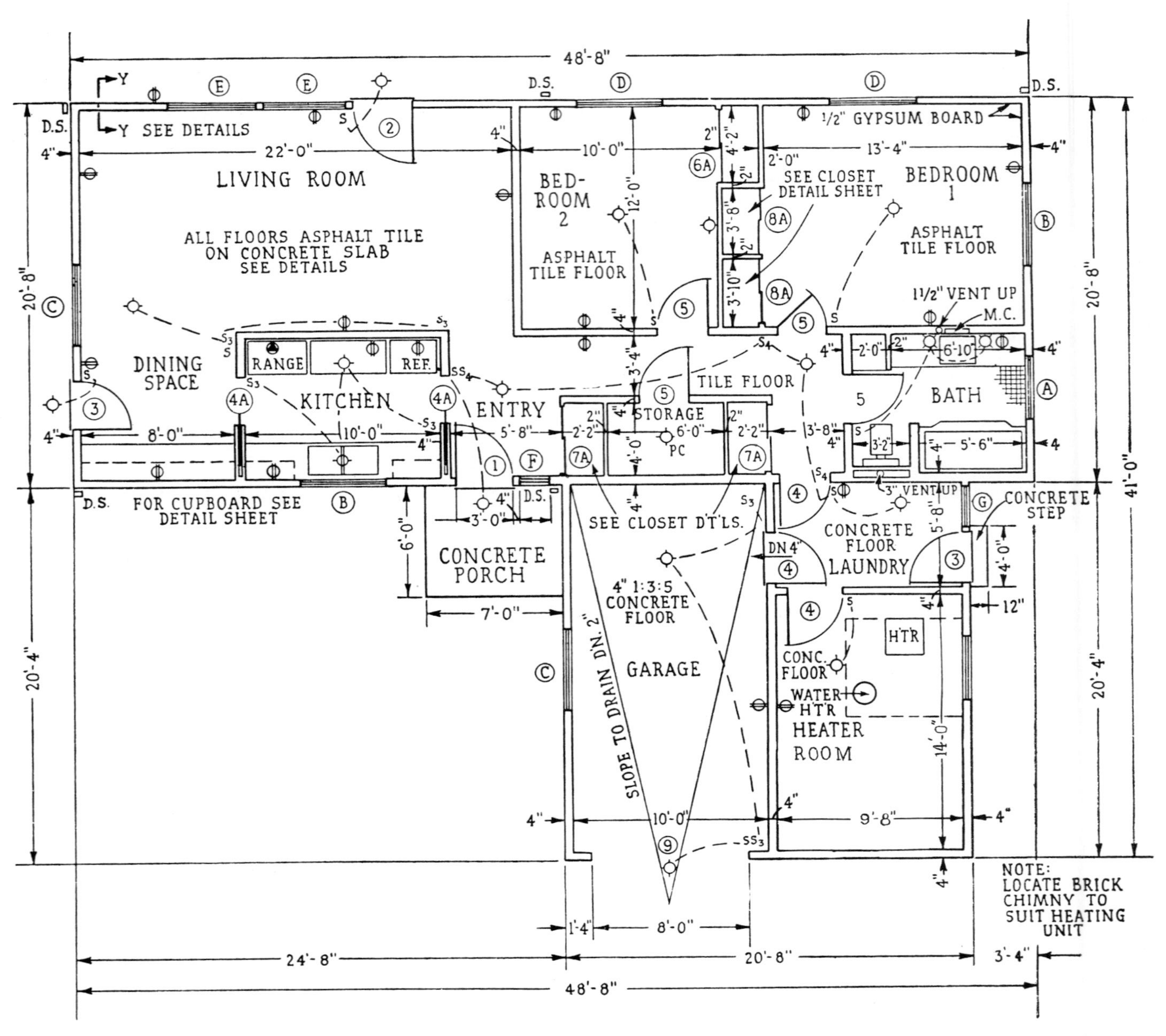

WINDOW SCHEDULE		
MARK	SIZE	TYPE
A	3'-4"x 3'-3"	HORIZONTAL GLIDING (SEE DETAIL SHEET)
B	4'-4"x 3'-3"	" " (" " ")
C	4'-4"x 3'-11"	" " (" " ")
D	4'-4"x 4'-7"	" " (" " ")
E	4'-4"x 5'-3"	" " (" " ")
F	16"x 84"	FIXED FIGURED GLASS (SEE DETAIL SHEET)
G	24"x 43"	" " " (" " ")

DOOR SCHEDULE		
MARK	SIZE	TYPE
1	13/4"x 3'-0"x 7'-0"	FLUSH PANEL (SEE DETAIL SHEET)
2	13/4"x 3'-0"x 7'-0"	PLATE GLASS PANEL 24"x 56"
3	13/4"x 2'-8"x 7'-0"	GLASS PANEL 22"x 38"
4	13/4"x 2'-8"x 6'-8"	FLUSH PANEL
4A	13/4"x 2'-8"x 6'-8"	FLUSH PANEL SLIDING DOOR
5	13/4"x 2'-6"x 6'-8"	FLUSH PANEL
6A	2 DOORS 13/8"x 2'-0"x 6'-8"	FLUSH PANEL SLIDING DOOR
7A	2 DOORS 13/8"x 1'-10"x 6'-8"	FLUSH PANEL SLIDING DOOR
8A	2 DOORS - 13/8"x 1'-10"x 6'-8"	FLUSH PANEL SLIDING DOOR
9	GARAGE DOOR - 8'-0"x 7'-0"	SEE ELEVATION (OVERHEAD DOOR)

FLOOR PLAN

SCALE 1/4" = 1'-0"

0 1 2 3 4 5 6 7 8 9 10 11 12 13 14

REDUCED SCALE

Figure 27A

windows. The next dimensions specify that the following three studs are to be spaced 1′ 4″ on center. The locations of other studs are specified in like manner.

Between the bathroom and bedroom 2, an 8-inch partition is required to provide space for the plumbing pipes. Note that the studs must have their sides parallel to the wall surfaces. Note, too, that a corner post, consisting of three studs, is necessary at the corner where the rear partition of the bathroom meets the partition for bedroom 2. Each corner of the chimney enclosure requires one-and-one-half studs.

The plan view of the utility level shows that two 2 by 10s are to be used as a header under the stairway landing. The left-hand end of that header is to be supported by a 6- by 6-inch post on a cast-in-place concrete base.

The closet for bedroom 1 is to be 2′ 2″ deep and 8′ 2″ long. Bedroom 2 is to have two closets, both to be 2′ 2″ deep and 4′ 0″ long.

The interior width of the garage is to be 10′ 0″ and its interior length is to be 21′ 4″. The length dimension is found by subtracting the thickness of two 8-inch concrete block walls from the 22′ 8″ overall dimension shown.

The ceiling joists over living and bedroom levels are to be 2 by 12s spaced 16 inches on center.

Electrical Each of the two bedrooms is to have ceiling lights controlled by wall switches. Each bedroom is also to have two wall convenience outlets.

The ceiling light in the sleeping-level hall is to be controlled by two switches, one near the door for bedroom 1, and one in the living-level hall. The garage ceiling light is also to be controlled by two switches.

Plumbing The bathroom is to have a 5′ 6″ bathtub, a water closet, and a lavatory.

Kitchen Equipment A note indicates that all kitchen equipment is shown on detail sheets.

Figures 27A and 27B. The following observations will be helpful in learning to read the plan views for a one-story house (Figure 12).

Materials The floor plan symbols show what various structural items are to be made of:

Exterior walls	*frame*	Garage floor	*concrete*
Bathroom floor	*tile*	Foundation	*concrete*
Chimney	*brick*	Interior partitions	*frame*
L. R. floor	*tile*	Laundry floor	*concrete*
Porch foundation	*concrete*	Floor slab	*concrete*

Details On several of the items, notes refer to detail sheets for necessary information. We will study details in a later chapter.

Windows The sizes and types of windows are shown in the schedule.

Doors All door sizes and types are shown in the door schedule. For example, door 3 must be 1¾ inches thick, 2′ 8″ wide, 7′ 0″ high, and contain a 22- by 38-inch glass panel. The door locations are indicated on a detail sheet.

Level All rooms except the heater room are on one level.

Garage Floor This floor is to be made of 4 inches of 1:3:5 cast-in-place concrete and is to slope 2 inches from back to front.

Main Floor The floor under all rooms is to be a 4-inch slab of cast-in-place concrete reinforced with 6- by 6-inch No. 10 wire mesh. The concrete is to be placed on a gravel fill. There must be a moisture barrier between the gravel and the concrete. In other words, the slab is to rest on the ground.

Foundation The notes indicate that foundation information will be found on detail sheets. The dashed line around the foundation indicates that the bottom of it is to be wider than the top. In other words, the wide bottom is to serve as a footing.

Chimney An approximate chimney location is indicated on the elevation views but no symbol is shown on the plan views. A note indicates that final position depends upon the heating unit to be selected.

General There is to be a concrete porch in connection with door 1.

The doors for the closets and kitchen must be the sliding type. The living room, dining space, and laundry must have exterior doors.

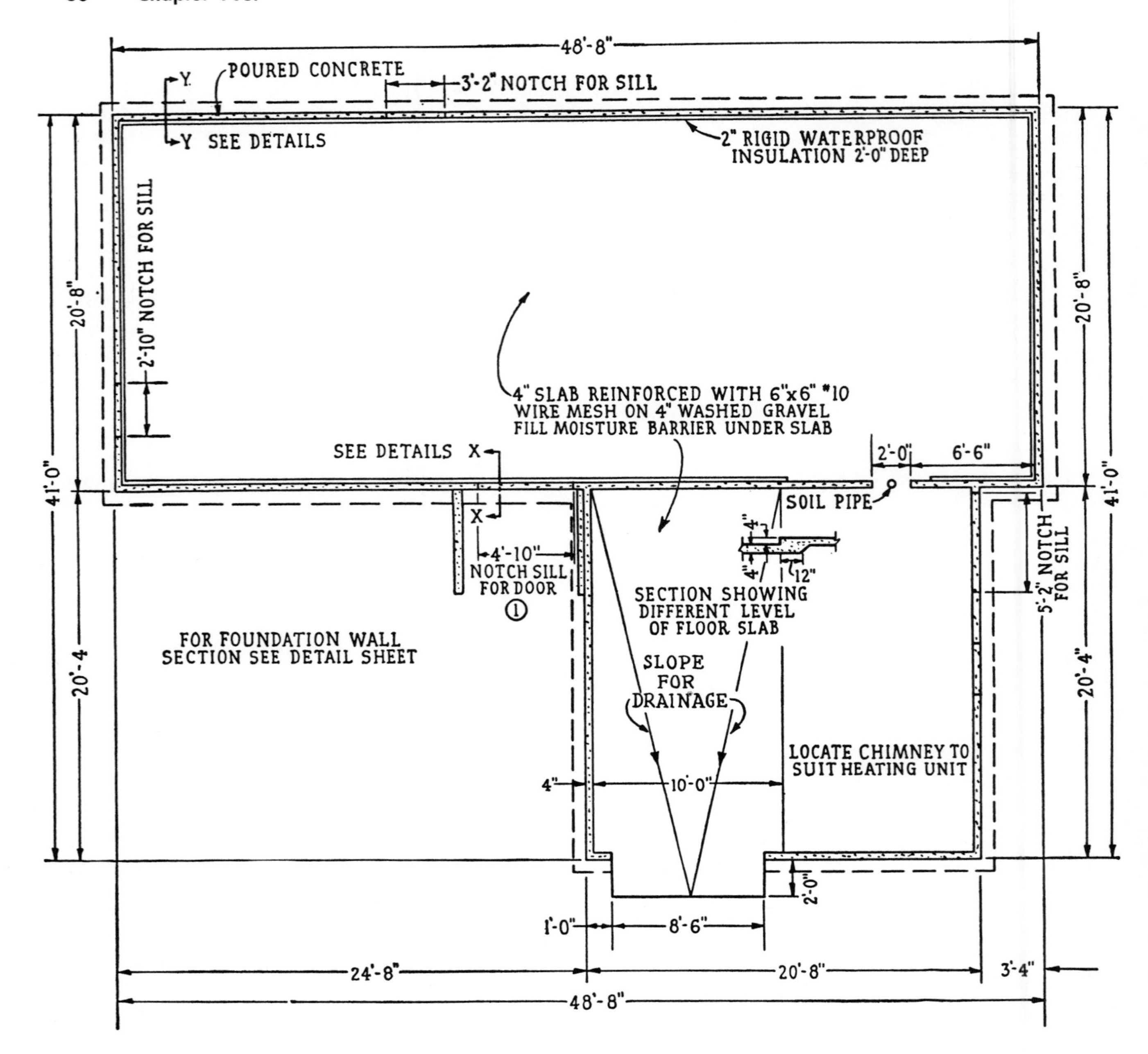

FOUNDATION PLAN

SCALE 1/4"=1'-0"

Figure 27B

The laundry must have a fixed window next to its exterior door.

Dimensions The greatest length of the house is to be 48′ 8″. The greatest width is to be 41′ 0″. The interior of the garage is to be 10′ 0″ wide and 20′ 0″ long. The interior of bedroom 1 is to be 13′ 4″ long and 12′ 0″ wide. The basement-floor level is to be 4 inches above the garage-floor level.

Electrical Switch-controlled ceiling lights are

required for the dining space, the kitchen, both bedrooms, the halls, laundry, heater room, and garage. A pull-chain light is specified for the storage room. In the bathroom, switch-controlled lights are to be located on both sides of the medicine cabinet.

The hall lights are to be controlled by three switches. The kitchen and garage lights and a living-room floor outlet are each controlled by two switches. A light is required outside of door 1.

There are to be six wall outlets in the combined dining space and living room. Each of the bedrooms is to have three wall outlets.

Plumbing The bathroom is to have a recessed bathtub, a recessed water closet, and a vanitylike lavatory. The soil stack is to be placed in an enlarged partition in back of the water closet.

Kitchen Equipment There is to be a range, a sink, a refrigerator, counters, and cabinets.

Figure 28. The following observations will be helpful in learning to read plan views for a two-story house (Figure 13).

Materials The plan and basement view symbols show that various structural items are to be made of:

Foundations	*concrete*	Areaway walls	*concrete*
Basement stairs	*concrete*	Cold closet	*frame*
Chimney	*brick*	Stairs	*wood*
Exterior steps	*flagstone*	Exterior walls	*frame*
Basement floor	*concrete*	Column footing	*concrete*
Chimney footing	*concrete*	Areaway floors	*gravel*
		Partitions	*frame*
Chimney piers	*brick*	Bookcases	*wood*
Rear steps	*concrete*	Counters	*wood*
Porch foundations	*concrete*	Railing	*wood*

Details The plan and basement views do not include detail notes. However, we can be sure that the plans contain ample details on other sheets.

Windows All window sizes are indicated in the elevation views.

Levels Even though the second-floor plan view is not shown in Figure 28, it is evident that the first-floor and basement views are part of the same house shown in Figure 13.

Stairs The staircase symbol on the first-floor plan shows that both the second-floor and basement stairs must have 12 risers. The basement plan view includes a symbol which indicates that a cast-in-place-concrete stairs must be provided which will allow access to the basement from the back yard. Eight risers are required.

General Two parallel beams are required to support floors and partitions above the basement. They are indicated by the long and short dashed lines. The two shorter beams are to be supported by the foundation and by piers which must be built as part of the chimney. The long beam must be supported along its length by two 4-inch Lally columns. The Lally columns are to be supported by cast-in-place-concrete footings. The ends of the long beam must be supported by the foundation.

There is to be a stone fill under the concrete steps required outside the kitchen door.

The cold closet is to have shelves on three sides.

Dimensions Those parts of the foundation which are to be under the brick-veneer end walls must be 12 inches thick. Other parts of the foundation are to be 10 inches thick.

The areaway walls are to be 6 inches thick, 5′ 0″ long (scaled), and extend out from the foundation 2′ 0″.

The interior of the basement is to be 31′ minus 2′ or 29′ 0″ long and 23′ 2″ minus 1′ 8″ or 21′ 6″ wide. Those dimensions had to be calculated because the thickness of the foundation had to be subtracted.

The interior dimensions of the basement exterior stairs show that a space 4′ 0″ wide and 4′ 0″ long is required.

The Lally columns must each be 7′ 2″ apart and 7′ 2″ from the foundation.

The overall dimensions show that the house must be 31′ 0″ long and 23′ 2″ wide.

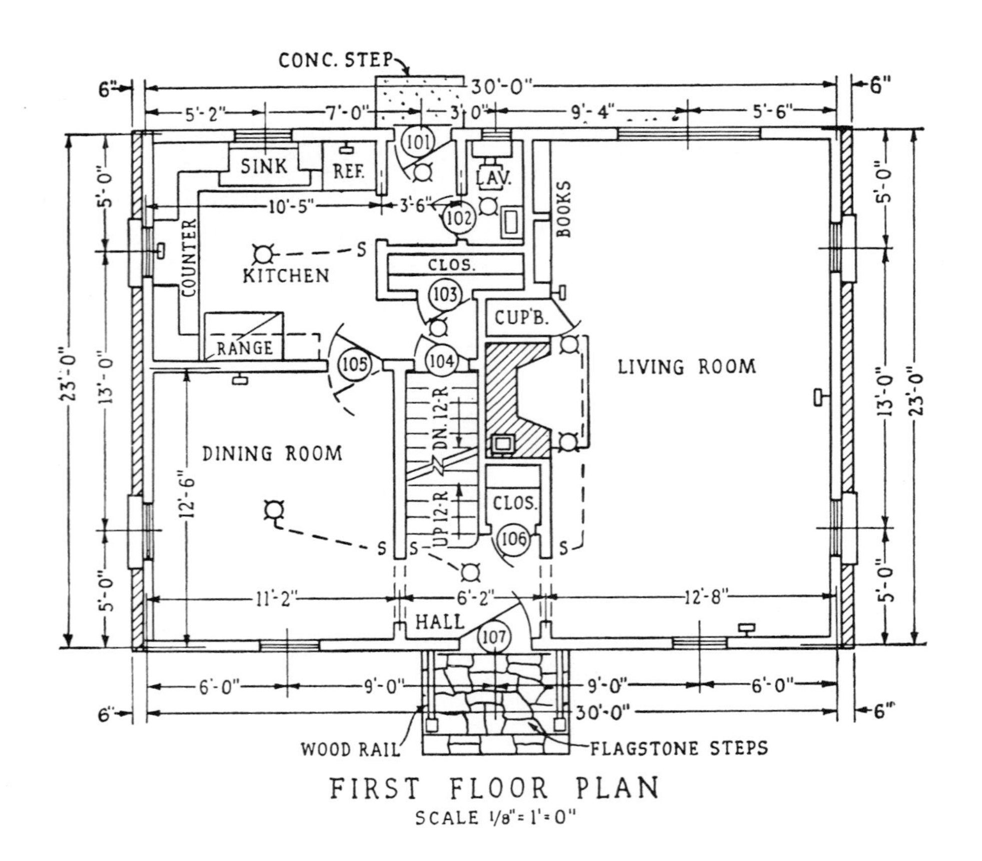

FIRST FLOOR PLAN

SCALE 1/8" = 1' = 0"

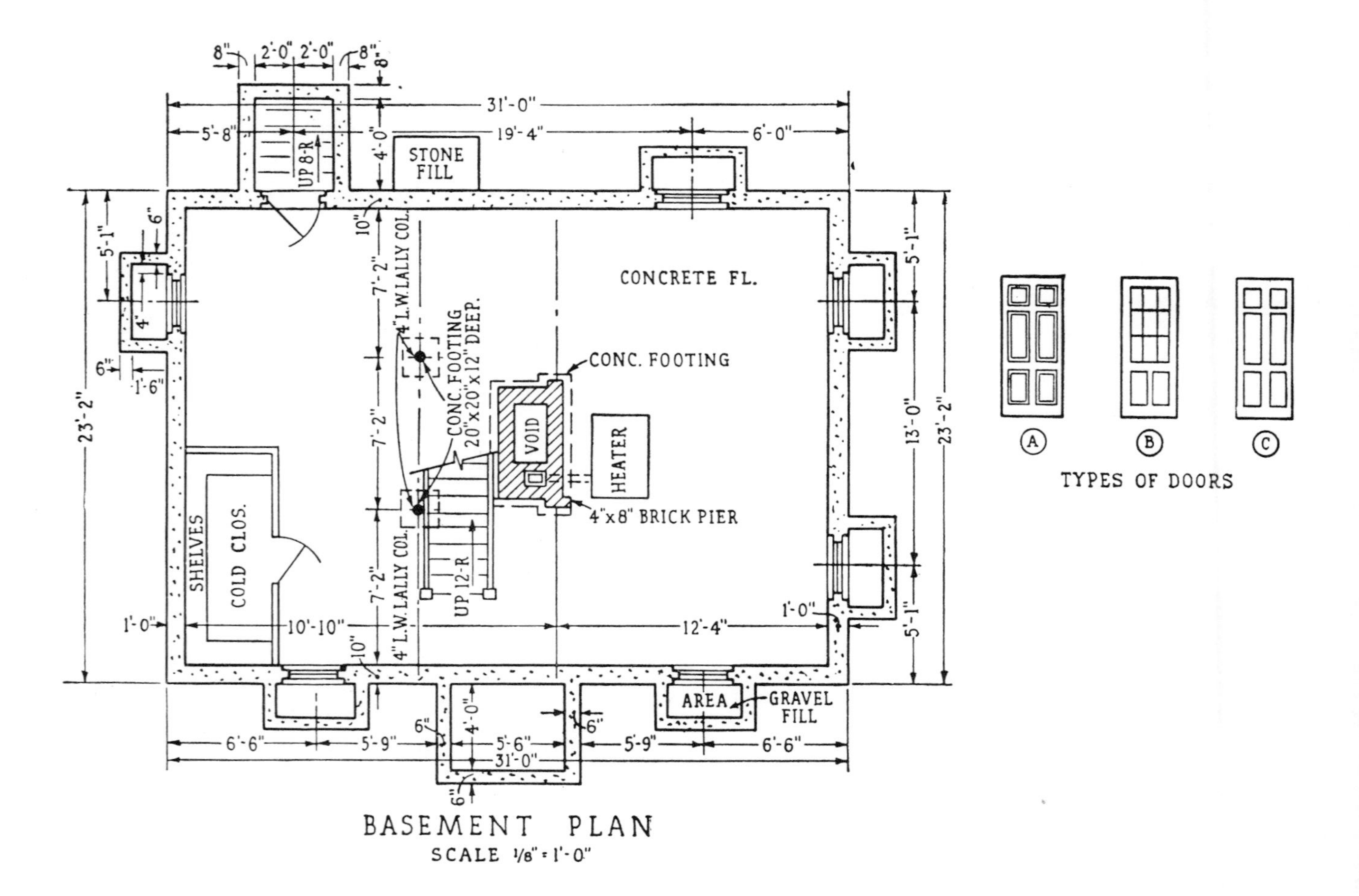

BASEMENT PLAN

SCALE 1/8" = 1'-0"

Figure 28

The center lines of the windows along the side of the living room must be located 5′ 0″ from the corners of the house. The center lines of the front windows in the living and dining rooms must be located 6′ 0″ from the corners.

Electrical The kitchen, dining room, and front hall must have ceiling lights, each of which is to be controlled by one switch. Three wall convenience outlets are required in the living room, one in the dining room, and two in the kitchen.

Sidewall lights are required on both sides of the fireplace and are to be controlled by one switch.

Plumbing There is to be one water closet and one lavatory for the first floor.

Kitchen Equipment A sink, a refrigerator, a range, cabinets, and counters are required.

QUESTIONS AND ANSWERS

The following questions and answers concern Figures 29 and 30 in this chapter, and Figures 14 and 15 in Chapter 3. All four of the illustrations represent the same house.

As a means of making sure that you have learned to visualize both plan and elevation views and that you understand the relationship between them, answer each of the questions, orally or in written form, and then check your reading with the *descriptive* answers shown. A review of this sort provides helpful practice and readers are urged to give the questions careful attention.

Question 1 According to the plan views, what is the shortest route from the kitchen to the back yard?
Answer 1 The shortest route is via the kitchen stairs to the den, through the den and laundry, and out through the laundry door.

Question 2 How many arched openings are indicated?
Answer 2 The symbols for two such openings are included in the symbols for the kitchen partitions.

Question 3 What are the interior dimensions of the closet in bedroom 2?
Answer 3 There are two dimensions within the closet symbol. They indicate that the interior of the closet is to be 2′ 0″ by 4′ 8″.

Question 4 How many risers are required for the stair between the kitchen and den levels?
Answer 4 The stair symbols, shown in both the den and kitchen areas, indicate that 7 risers are required.

Question 5 What tread width is necessary for the stairs mentioned in the previous question?
Answer 5 The tread-width dimension is not shown. However, the scaling process shows that the width is to be 8 inches.

Question 6 How thick are the foundations to be and what material is to be used?
Answer 6 The symbols indicate that two thicknesses are involved. On the ground-floor plan, the thickness of the foundation under the living-room bay windows is specified as 10 inches. All other parts are specified as 8 inches thick. Two materials are indicated. Concrete block is required for that portion of the foundation to the left of the den and laundry areas. Cast-in-place concrete is required for all other portions.

Question 7 Is concrete block to be used anywhere other than for a portion of the foundation?
Answer 7 The symbols in the ground-floor plan indicate that a concrete-block curb is required under the garage partition.

Question 8 How much of the area under the house is to be excavated?
Answer 8 Excavation is required only for the den, laundry, heater room, and garage areas.

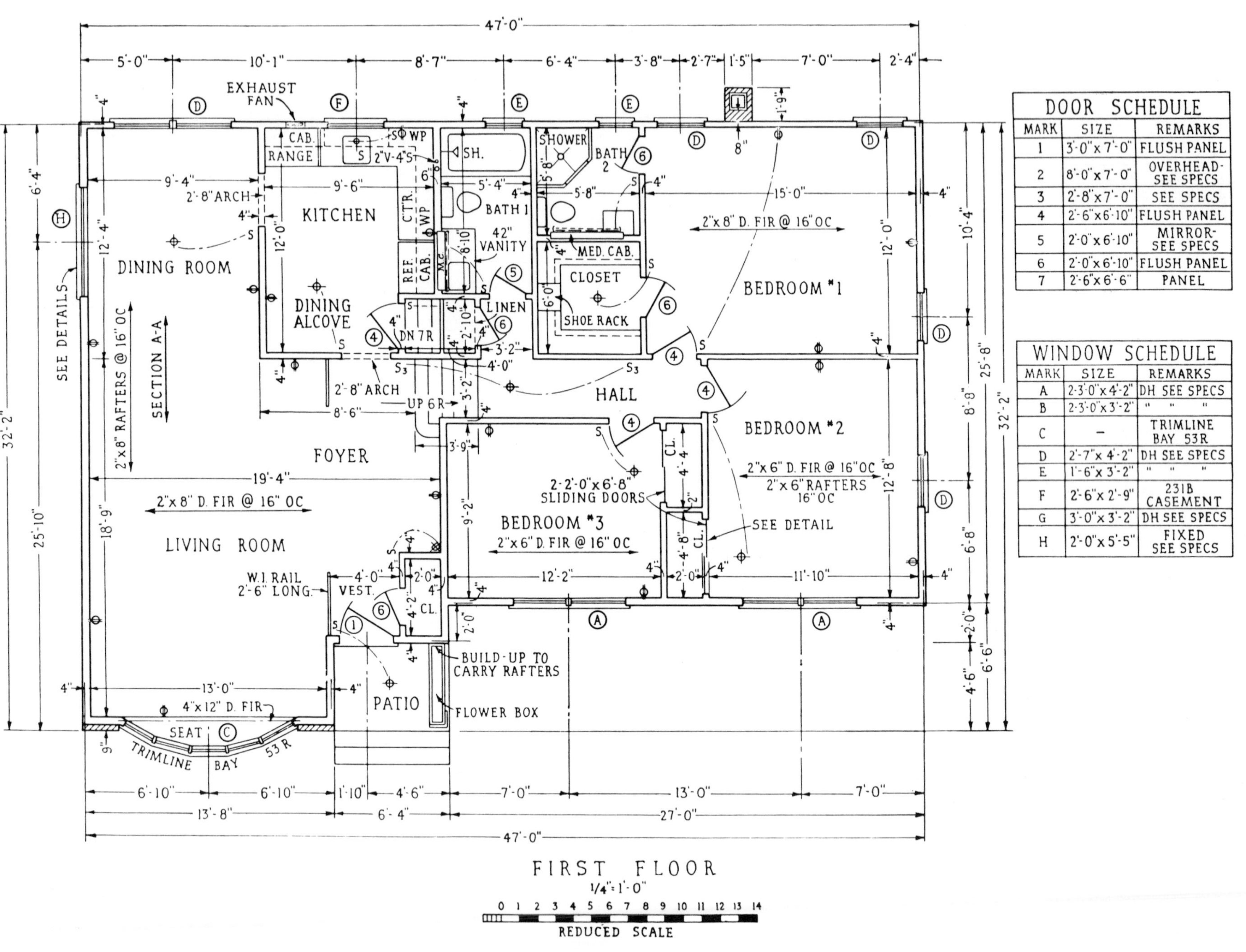

DOOR SCHEDULE		
MARK	SIZE	REMARKS
1	3'-0"x 7'-0"	FLUSH PANEL
2	8'-0"x 7'-0"	OVERHEAD-SEE SPECS
3	2'-8"x 7'-0"	SEE SPECS
4	2'-6"x 6'-10"	FLUSH PANEL
5	2'-0"x 6'-10"	MIRROR-SEE SPECS
6	2'-0"x 6'-10"	FLUSH PANEL
7	2'-6"x 6'-6"	PANEL

WINDOW SCHEDULE		
MARK	SIZE	REMARKS
A	2-3'-0"x 4'-2"	DH SEE SPECS
B	2-3'-0"x 3'-2"	" " "
C	–	TRIMLINE BAY 53R
D	2'-7"x 4'-2"	DH SEE SPECS
E	1'-6"x 3'-2"	" " "
F	2'-6"x 2'-9"	231B CASEMENT
G	3'-0"x 3'-2"	DH SEE SPECS
H	2'-0"x 5'-5"	FIXED SEE SPECS

FIRST FLOOR

1/4" = 1'-0"

Figure 29

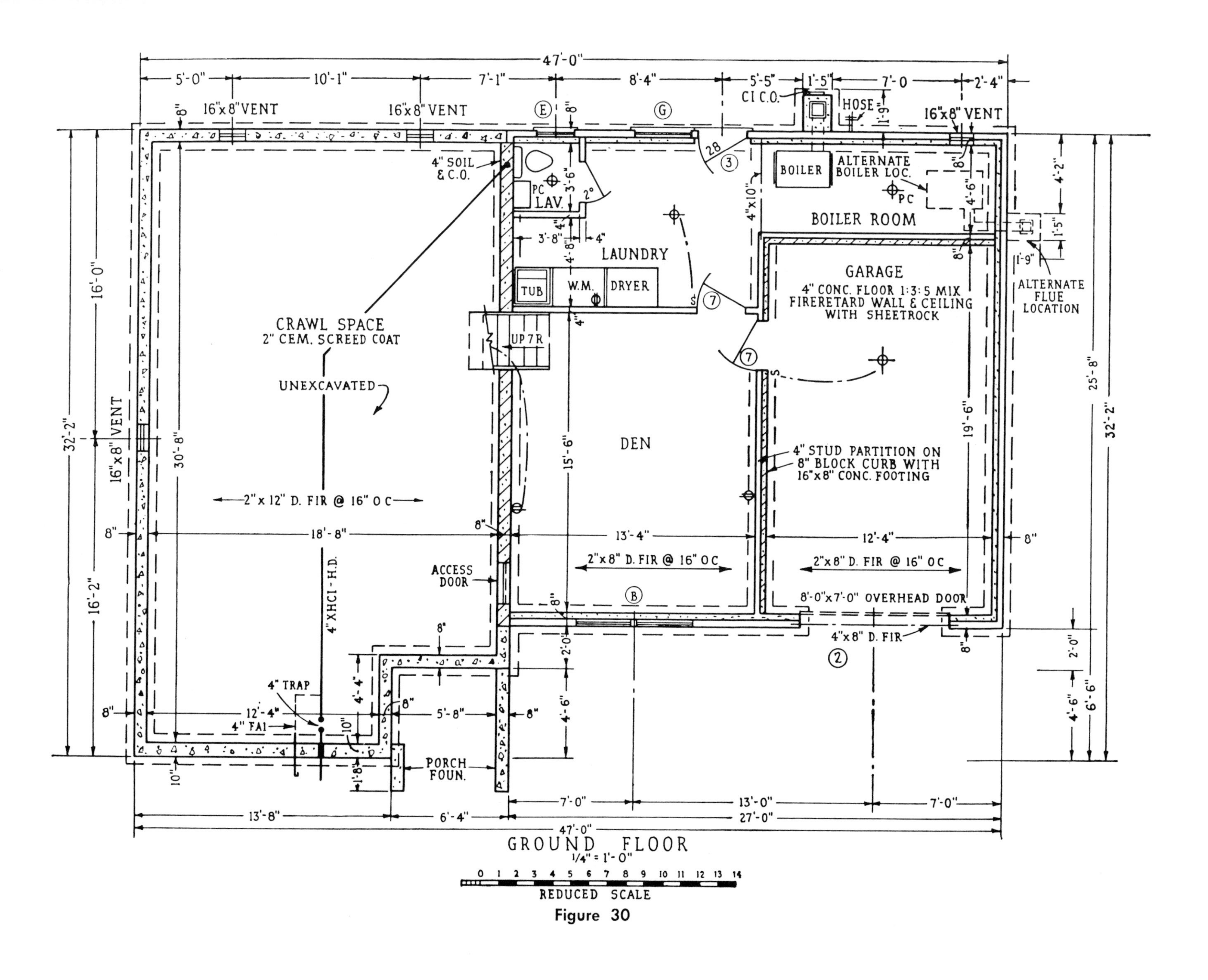
CRAWL SPACE
2" CEM. SCREED COAT
UNEXCAVATED
2"x 12" D. FIR @ 16" O C
4" SOIL & C.O.
4" XHCI - H.D.
4" TRAP
4" FAI
ACCESS DOOR
PORCH FOUN.
LAV.
PC
LAUNDRY
TUB
W.M.
DRYER
UP 7R
DEN
2"x8" D. FIR @ 16" O C
BOILER
ALTERNATE BOILER LOC.
BOILER ROOM
CI C.O.
HOSE
GARAGE
4" CONC. FLOOR 1:3:5 MIX
FIRERETARD WALL & CEILING
WITH SHEETROCK
4" STUD PARTITION ON
8" BLOCK CURB WITH
16"x8" CONC. FOOTING
8'-0"x7'-0" OVERHEAD DOOR
4"x8" D. FIR
ALTERNATE FLUE LOCATION
16"x8" VENT
47'-0"
32'-2"
GROUND FLOOR
1/4" = 1'-0"
REDUCED SCALE

Figure 30

Question 9 What are the interior dimensions of the crawl space to be?
Answer 9 Two dimensions indicate that the space is to be 30′ 8″ long and 18′ 8″ wide.

Question 10 What kind of joists are required for the living-room and kitchen floors?
Answer 10 There is a symbol in the crawl-space area which indicates that the joists must be made of Douglas fir.

Question 11 What size and spacing are required for the joists mentioned in the previous question?
Answer 11 The symbol indicates that 2 by 12 joists are required and that they must be spaced 16 inches on centers.

Question 12 If joists are obtained in lengths which are multiples of 2, what length would you buy for the living-room and kitchen floors?
Answer 12 In this case, 20-foot lengths would be required because the span is 18′ 8″ and the joists must have sufficient bearing at each end.

Question 13 What size house drain is required?
Answer 13 In the crawl-space area, the house drain size is specified as 4 inches.

Question 14 What size door is required for bedroom 2?
Answer 14 The letter 4, near the symbol, refers to the door schedule which shows that the door must be 2′ 6″ by 6′ 10″.

Question 15 How many windows are required for the living-room bay?
Answer 15 The symbol in the first-floor plan shows 5 windows.

Question 16 What kind of doors are required for the closets in bedrooms 2 and 3?
Answer 16 The symbols indicate sliding doors.

Question 17 How many medicine cabinets are required?
Answer 17 Medicine cabinet symbols are included with the symbols for both bathrooms.

Question 18 What is the run for the stairs between the living-room and bedroom levels?
Answer 18 In the stair symbol, which is shown on the first-floor plan, the run is specified as 3′ 9″.

Question 19 How many dining-room areas are indicated?
Answer 19 On the first-floor plan, two are indicated: one in connection with the living room and one in connection with the kitchen.

Question 20 What material is to be used for the patio floor?
Answer 20 The first-floor plan does not show a symbol for that floor. However, the front elevation indicates that the floor is to be made of cast-in-place concrete.

Question 21 What is the exact position of the den windows?
Answer 21 The front elevation shows that the heads of the windows are 6′ 5″ above the grade. The ground-floor plan shows that the center line between the two windows is 7′ 0″ + 13′ 0″ or 20′ 0″ from the right-hand corner of the garage.

Question 22 What is the exact position of the kitchen window?
Answer 22 The rear elevation shows that the head is 7′ 6″ above the top of the foundation. The first-floor plan shows that the center line of the window is 5′ 0″ + 10′ 1″ or 15′ 1″ from the corner of the dining room.

Question 23 What size rafters are required in the roof over the bedrooms?
Answer 23 In bedroom 2 there is a symbol which indicates that 2 by 6 rafters are required.

Question 24 Are any large-sized wood timbers required?
Answer 24 The first-floor plan shows a 4 by 12 timber over the living-room bay windows. The ground-floor plan shows that a 4 by 10 timber is required in the laundry and heater-room area and that a 4 by 8 timber is required over the garage door.

Question 25 Of what materials are the walls and partitions to be made?
Answer 25 All walls and partitions are indicated as being of frame, or wood, construction, except the front wall of the living room which is indicated as brick veneer.

Question 26 What equipment is indicated in the kitchen?
Answer 26 The symbols indicate a refrigerator, a sink, a range, cabinets, and an exhaust fan.

Question 27 On how many sides of the kitchen are there cabinets?
Answer 27 The cabinets extend around two sides.

Question 28 What are the interior dimensions of the laundry?
Answer 28 There is a 13′ 4″ dimension in the den area which also applies to the laundry. The width must be found by adding the short dimensions which go through the lavatory area. Thus, 4′ 8″ + 4″ + 3′ 6″ = 8′ 6″. Note that the 4-inch dimension for the partition thickness had to be included.

Question 29 How many electrical wall outlets are indicated for the living and dining rooms?
Answer 29 There are four symbols.

Question 30 How many electrical ceiling outlets or fixtures are indicated in the first-floor plan?
Answer 30 The dining room has one. The kitchen has two, including the one over the sink. The hall has one. The closet for bedroom 1 has one. Each of the three bedrooms has one. The patio has one.

Question 31 What type of lights are indicated for the bathrooms?
Answer 31 The lights are indicated as fluorescent and are placed over the medicine cabinets.

Question 32 How many water closets are required for the house?
Answer 32 One is required for each of the two bathrooms and one for the lavatory.

Question 33 Are any of the partitions to be thicker than four inches?
Answer 33 The partition between the kitchen and bathroom is to be 6 inches thick so that the soil stack can be placed within it.

Question 34 What is the size of the *A* windows?
Answer 34 The window schedule indicates that the *A* windows must be 2′ 3″ by 4′ 2″.

Question 35 Are any shelves indicated?
Answer 35 Shelves are indicated on three sides of the closet for bedroom 1 and in the linen closet.

Question 36 What size are the linen closet shelves to be?
Answer 36 The dashed line, near the symbol, means that the shelves are to be the size of the interior of the closet. The dimensions indicate that the interior of the closet is 2′ 10″ long. No width is given but it scales as 1′ 9″.

Question 37 What size soil stack is required?
Answer 37 Near the lavatory symbol, in the ground-floor plan, the size is specified as 4 inches.

Question 38 How many pull-chain lights are indicated?
Answer 38 One in the lavatory and one in the heater room.

Question 39 What size of joists are required for the ceiling over the living room?
Answer 39 A symbol, in the living-room area, shows that 2 by 8 joists are required.

Question 40 What thickness is necessary for the brick-veneer wall?
Answer 40 The dimension just to the left of the bay-window symbol on the first-floor plan indicates 9 inches.

Question 41 How is the foyer to be lighted?
Answer 41 The first-floor plan shows the symbol for a floodlight. This is an uncommon symbol. On the floor plan, it is located in the corner where the rear partition of the entrance closet meets the partition between the foyer and bedroom 3.

Question 42 From your study of the plan and elevation views, can you tell what type of fuel is to be used for heating purposes?
Answer 42 There is no indication of a coal bin. Thus, we can reason that either gas or oil will be used as fuel. The small flue in the chimney also indicates that solid fuels are not to be burned.

Question 43 What do the ground-floor symbols tell us about the garage partition?
Answer 43 The footing under the partition is to be 8 by 16 inches and of cast-in-place concrete. One row of 8-inch concrete block is to rest on the footing, as a curb. A stud partition is to extend

from the top of the curb to the ceiling. Sheetrock is to be used as a means of fireproofing.

Question 44 What material is to be used for the garage floor?
Answer 44 A note, in the garage area, states that 4 inches of concrete shall be used to make the garage floor. The note also states that the concrete shall be a 1 : 3 : 5 mix. That means 1 part of cement, 3 parts of sand, and 5 parts of crushed rock.

Question 45 Do the symbols indicate that any portion of the foundation does not require footings?
Answer 45 The ground-floor plan indicates that footings are not required for the front porch foundations.

Question 46 How would you determine the length of the longer of the two portions of porch foundations?
Answer 46 The ground-floor plans show two dimensions which can be added to find the length of that portion of porch foundation. Note the 4′ 6″ dimension to the right of the foundation. Also, note the 1′ 8″ dimension to the left of the smaller portion of the porch foundation. The sum of these two dimensions indicates the length required.

Question 47 How many bathtubs are required?
Answer 47 Only bathroom 1 requires a bathtub. Bathroom 2 is to have a shower enclosure in place of a bathtub.

Question 48 Would you say that the foyer-floor level is at the level of the porch floor?
Answer 48 The front elevation shows that there is to be a concrete sill under the door marked 1. This indicates that the foyer is to be one step higher than the porch floor.

Question 49 How many vents are required for the crawl space?
Answer 49 Three vent symbols are shown on the ground-floor plan.

Question 50 What size are the crawl-space vents to be?
Answer 50 The rear elevation shows that the vents are to be 8 by 16 inches.

CHAPTER FIVE

Section Views

In many instances, elevation and plan views cannot show sufficient information to enable carpenters and other mechanics to see exactly how the various structural parts of a structure are to be built or assembled. For example, the elevation views, in Figure 12, show that the exterior walls of the structure are to be of frame construction. This means that, unless otherwise directed in the specs, they would be constructed using 2 by 4 studs, with lath and plaster on the inside, and sheathing, building paper, and some sort of siding on the outside. When we studied Figures 11 and 26*A*, we found that gypsumboard was specified for the interior finish and wood siding for the exterior finish. We also found that exact locations for studs are indicated. No other information is shown.

There are several good and accepted ways of constructing frame walls, any one of which can be used to advantage. There are manifold types of sill construction, various methods of bracing, and different constructions for other items which are not shown in Figures 11 and 26*A*. Thus, additional drawings, which we shall call *section views,* are necessary.

Section views are also picturelike representations in which symbols, terms, and abbreviations are used as a means of indicating a great deal of information. In addition, section views show how various structural items are to be assembled or built and how they are related to each other in the general construction of a proposed structure.

The purpose of this chapter is to show why section views are necessary, how to visualize them, and how to read or use them in connection with elevation and plan views. We shall be concerned with the fundamentals of section views rather than their application to popular types of structures. Once we understand these fundamentals, applications will be clear.

WHY ARE SECTION VIEWS NECESSARY?

We might ask why section views are necessary and wonder why not enough information can be shown on elevation and plan views to satisfy the need.

There are two basic reasons why section views are necessary. First, elevation and plan views cannot be drawn to a large enough scale to show section view items clearly. This fact will be understood as we study typical section views later in this chapter. Second, neither elevation nor plan views can be drawn in a manner to include section views. The following examples will help us to understand why section views are necessary.

Staircases When we studied the interior staircase symbols shown in Figures 26*A* through 30, we learned where they are to be located, the required number of risers, how wide they are to be, and their general type. But we did not find any information which indicated the actual construction. In order to build such staircases, a carpenter must know how many stringers to use and their size, the thickness of the treads and risers, how the treads and risers are to be joined to the stringers, how the stringers are to be attached to the framing, and how the railing and newel posts are to be made. Such information cannot be shown on elevation or plan views. Therefore, additional drawings are necessary.

Fireplaces When we studied Figure 28 we found some information about the fireplace. The symbol indicated that it was to be built of brick. But from the plan views we know nothing about the height of the fireplace, how the trimmer arch is to be constructed, how the brickwork over the fireplace opening is to be supported, or how the interior is to be built. Here again, some additional drawings are necessary.

Steel Beams Figure 26*B* notes that a W8 beam is necessary to support the living-area level over the crawl space. But there are no indications as to how the floor joists are to be fastened to the beam or how the beam ends are to be connected to the supporting pilasters. Additional drawings are needed.

In our study of elevation and plan views we saw many other examples of the need for such additional drawings. In fact, many of the notes refer to such additional drawings as we shall study in this chapter.

HOW TO VISUALIZE SECTION VIEWS

Section views are often included with, or spoken of as, *detail drawings*. In fact, many drawings include both section and detail views under the general title of structural details. For our purpose, and as a means of easier and better visualization, we shall consider section and detail views separately and in separate chapters. We must understand section views before we can visualize detail views.

Strictly speaking, section views indicate the interiors and arrangements of individual or related structural parts, whereas detail views make use of section views to show enlarged "pictures" of items too small to be represented on the scales used for elevation and plan views. Detail views also show elevation views of room interiors.

It is not necessary for mechanics to be acquainted with strength of materials or other engineering subjects, in order to read and follow the directions shown in section views. In other words, architects and engineers do the design work and mechanics do the building work.

Section views, like plan views, are carefully and accurately drawn to scale. However, much larger scales are generally used for section views.

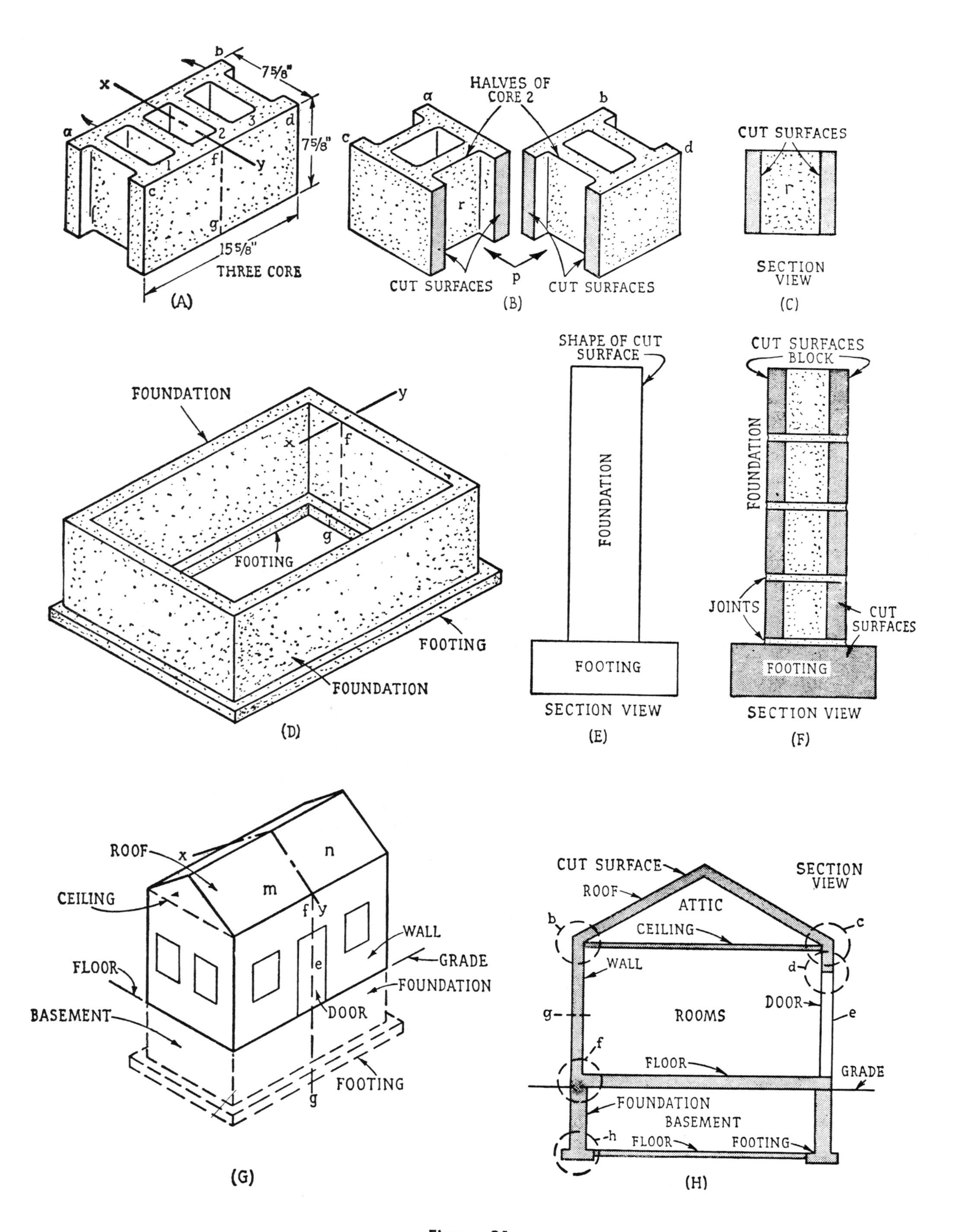

b
7 5/8"
x
a
2
3
d
7 5/8"
1
f
y
c
g
15 5/8"
THREE CORE
(A)
HALVES OF
CORE 2
a
b
c
d
r
CUT SURFACES
p
CUT SURFACES
(B)
CUT SURFACES
r
SECTION
VIEW
(C)
FOUNDATION
y
x
f
g
FOOTING
FOOTING
FOUNDATION
(D)
SHAPE OF CUT
SURFACE
FOUNDATION
FOOTING
SECTION VIEW
(E)
CUT SURFACES
BLOCK
FOUNDATION
JOINTS
CUT
SURFACES
FOOTING
SECTION VIEW
(F)
ROOF
x
n
m
CEILING
f
y
WALL
GRADE
FLOOR
e
FOUNDATION
BASEMENT
DOOR
FOOTING
g
(G)
CUT SURFACE
SECTION
VIEW
ROOF
ATTIC
b
c
CEILING
WALL
d
DOOR
g
ROOMS
e
f
FLOOR
GRADE
FOUNDATION
BASEMENT
h
FLOOR
FOOTING
(H)

Figure 31

When we learned to visualize plan views, we imagined that a complete structure could be cut through *horizontally* so that the top part could be moved away, exposing the cut surface of the bottom part. This was explained in connection with Figure 16. In order to visualize section views, we must imagine that complete structures and structural parts of them can be cut through *vertically*. The following examples explain the visualization process.

Examples The *A* part of Figure 31 shows a pictorial view of an ordinary concrete block which has three cores. Let us imagine that we can use a saw to cut the block *vertically* in half, as indicated by the cutting line *xy*. The dashed line *fg* indicates that the saw cuts the block vertically. Next, imagine that, after the cutting operation, we can swing the *a* and *b* corners backward in the direction of the arrows. The *B* part of the illustration shows the two halves of the block after the two corners have been moved backward so as to open up the cut surfaces. The cut surfaces are shown in black. Note that core 2 has also been cut in half. If we imagine that our eyes are at the position of the point *p*, we can look in two directions, as indicated by the arrows, and see two cut surfaces. The *C* part of the illustration shows what constitutes a section view of the concrete block.

The *D* part of Figure 31 shows a pictorial view of such concrete footings and foundations as might be used under a building. The surrounding earth has been omitted so that the footings and foundation can be seen. Note that the footing, as we learned in our study of elevation and plan views, is wider than the foundation is thick. Let us imagine that we can make a *vertical* cut through the footings and foundations as indicated by the cutting line *xy*. The dashed line *fg* shows the path of the saw in making the cut. Part *E* of the illustration shows the shape of the cut surface. If we imagine that the foundations are made of concrete block, the *F* part of the illustration shows the section view.

The *G* part of the illustration shows a houselike sketch with footings and foundations under it. Let us imagine that we can cut the structure and basement *vertically* in half at the point indicated by the cutting line *xy*. The dashed line *fg* shows the path of the saw. Note that the cut goes through the door opening. Next, let us imagine that the *m* part of the structure can be moved away so we can look at the cut surface of the *n* part. The *H* part of the illustration shows the section view. Note that the doorway is not shown in section because there was nothing to cut.

As in Chapter 4, we shall indicate cut surfaces in black until such time as we discuss section symbols.

The *A* part of Figure 32 shows a pictorial sketch of one of the many ways in which the rough framing for a house can be erected. We need not consider all the other framing methods because sections of them are visualized in exactly the same manner as we shall discuss in the following.

Readers are urged to study the framing sections as a means of becoming familiar with the terms and general locations of the various framing members.

As previously mentioned, elevation and plan views cannot show how the framing for an exterior wall is to be assembled or built. For example, note the parts of the section marked *b*, *g*, and *f* in the *H* part of Figure 31. Carpenters have to know how such parts are to be built. To show the required information, section views are needed.

The *B* part of Figure 32 shows a typical cornice section view. This view represents what would be seen if we could cut the framing in the *A* part of the sketch along the cutting line *bb*. Only the plate would be cut. Thus, the section view shows a cut surface for only the plate. The rafter, joist, and stud simply constitute a flat view of these members.

The *C* part of the illustration shows a section view such as we would see if we cut through

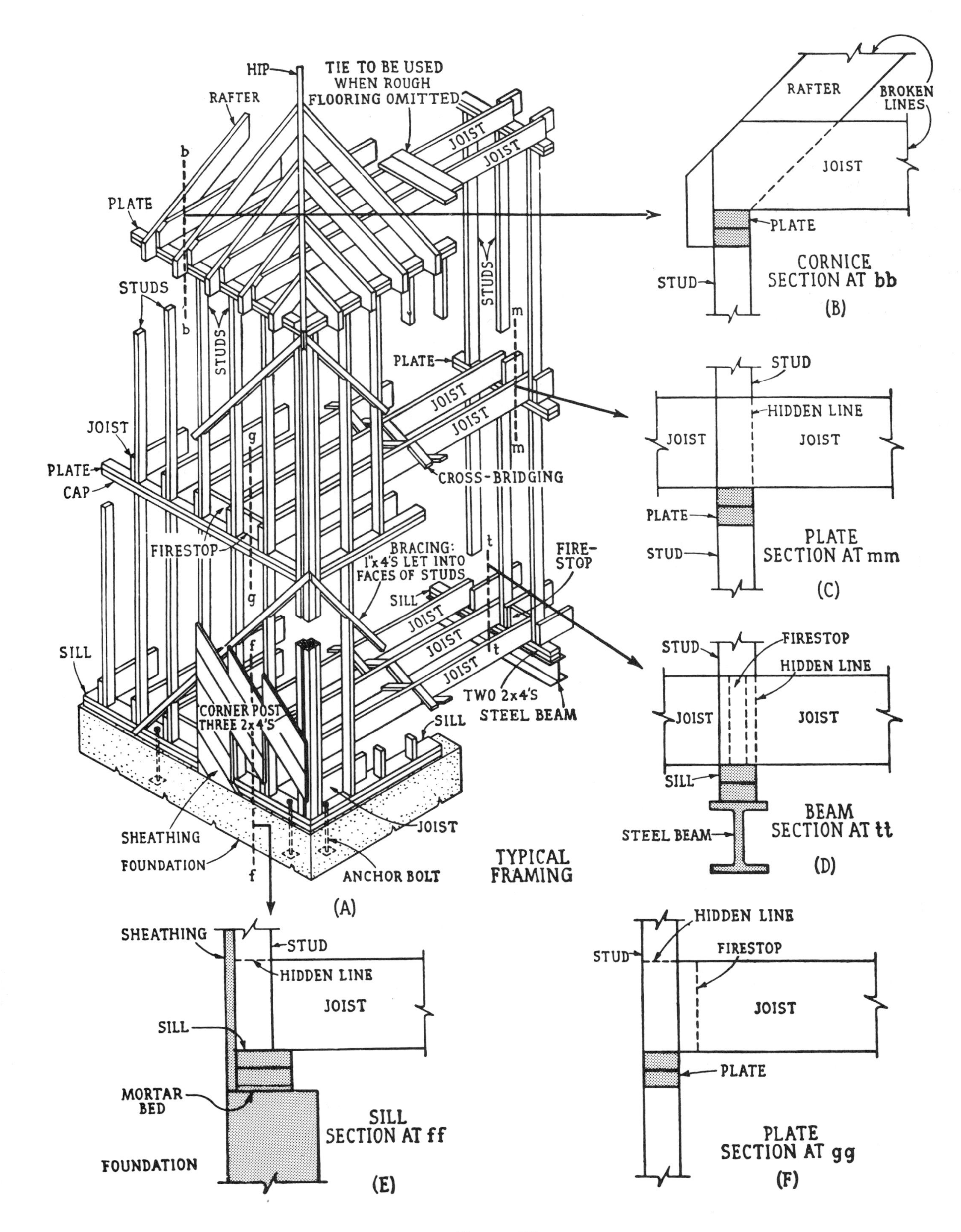
HIP
TIE TO BE USED WHEN ROUGH FLOORING OMITTED
RAFTER
JOIST
JOIST
b
PLATE
STUDS
b
STUDS
STUDS
m
PLATE
JOIST
JOIST
m
JOIST
PLATE
CAP
g
CROSS-BRIDGING
FIRESTOP
BRACING: 1"x4'S LET INTO FACES OF STUDS
t
FIRE-STOP
g
SILL
JOIST
f
JOIST
t
SILL
JOIST
CORNER POST THREE 2x4'S
TWO 2x4'S
STEEL BEAM
SILL
SHEATHING
JOIST
FOUNDATION
f
ANCHOR BOLT
TYPICAL FRAMING
(A)
RAFTER
BROKEN LINES
JOIST
PLATE
CORNICE SECTION AT bb
STUD
(B)
STUD
HIDDEN LINE
JOIST
JOIST
PLATE
STUD
PLATE SECTION AT mm
(C)
STUD
FIRESTOP
HIDDEN LINE
JOIST
JOIST
SILL
BEAM SECTION AT tt
STEEL BEAM
(D)
SHEATHING
STUD
HIDDEN LINE
JOIST
SILL
MORTAR BED
SILL SECTION AT ff
FOUNDATION
(E)
HIDDEN LINE
STUD
FIRESTOP
JOIST
PLATE
PLATE SECTION AT gg
(F)

Figure 32

the plate on top of the interior partition along the cutting line *mm*. Here again, only the plate is shown with a cut surface because the joists and stud were not cut. The dashed line indicates one edge of the stud which is hidden behind the joist.

The *D* part of the illustration shows a section view through the beam that supports the interior partition, as though it had been cut along the cutting line *tt*. In this case, the sill and the beam are shown with cut surfaces because they were the parts cut. The dashed lines indicate the fire stop between the joists.

Note that the broken lines, which we learned about in Chapter 4, are used to indicate that only portions of the rafters, joists, and studs are shown.

The *E* part of the illustration shows a section view of the part indicated by the dashed-line circle, marked *f*, in the *H* part of Figure 31. The section view also refers to the cutting line *ff* in the *A* part of Figure 32. Three different cut surfaces are shown: one for the sill, one for the sheathing, and one for the foundation. The dashed line indicates that part of the joist is hidden by the stud.

The *F* part of the illustration shows a section view as though the exterior plate were cut along the line *gg*. Only the plate is shown as a cut surface because the other members were not cut. The horizontal dashed line indicates the top edge of the joist which is hidden by the stud. The vertical dashed line indicates a hidden edge of the fire stop.

Section views of other structural members and assemblies, including whole structures, are visualized in exactly the same manner. Such views, unlike plan views, are always *vertical* and are generally drawn using a much larger scale than for either elevation or plan views.

Some architects elect to show only elevation and plan views on the drawings having these names. Then, as shown in Figures 11, 12, 13, and 26*A* through 30, they use notes to indicate that *details* are shown on other sheets of the drawings. The term *detail* includes both section and detail views. Or, they may make no mention of detail sheets with the understanding that drawing readers will know such sheets are included with the house plans. Other designers prefer to use somewhat larger print sheets so that they can include section and detail views along with the elevation and plan views.

RELATIONSHIP BETWEEN SECTION VIEWS AND PLAN AND ELEVATION VIEWS

At this stage in our study of section views, we can readily understand that, as shown in Figure 32, they cannot be drawn as parts of either elevation or plan views. Elevation views show only the *exterior appearances* of windows, doors, walls, roofs, and other parts of houses. While such views do contain much needed information, they cannot show anything about the interiors. Plan views show only *horizontal* cuts which indicate the interior arrangements of structures. And, even more than in elevation views, they include a great array of absolutely necessary information. But they cannot show how the structural members are to be assembled or built. Thus, as explained earlier in this chapter, section views are necessary.

In our study of Figures 31 and 32 we learned that in order to visualize section views, we must imagine that various structural members, or groups of them, can be cut through *vertically* and that after the cutting the members can be moved apart to expose the cut surfaces. We learned, too, that such cut surfaces show what we call section views.

As a means of visualizing the relationship between section views and elevation and plan views, let us review Figures 14 and 15 along with Figures 29 and 30. The elevation and plan views in these illustrations provide us with a great deal of information concerning the structure they represent. Yet a great deal more information actually is needed.

Example The plan view in Figure 29 includes a double-headed arrow along the symbols which represent the living room. The arrow is labeled "Section *A-A*." Note that the arrow is parallel to the 32′ 2″ overall dimension. One head of the arrow points to the bay-window symbol and the other to the *D* window symbols. The arrow means that we must imagine that the whole structure, from the top of the roof down to the bottom of the footings, can be cut through along a line which coincides with the arrow and extends from front to back of the structure. We must also imagine that after the cutting, the structure can be moved apart to expose the cut surface.

Figure 33 shows the section view we would see if we looked at the cut surface. Note the bay and *D* windows. They will help us to visualize the position of the section view in relation to the elevation and plan views. Note, too, that in Figure 30 the floor-joist symbol is shown at right angles to the direction of the double-headed arrow in Figure 29. In Figure 33, the floor joists (a representative few are shown) are at right angles to the cut surface. In like manner, the ceiling joists are also at right angles to the cut surface.

Note the 8′ 1″ and the 6′ 6″ dimensions shown at the left side of Figure 14. These same dimensions are shown at the left side of the section view and serve to clarify our visualization of the ceiling and window-head heights above the living-room floor.

The section view shows us the following information, none of which could have been shown on elevation or plan views.

1. The sill construction under the dining-room exterior wall must use two 2 by 6s as a sill, have a 2 by 10 header, and a 2 by 4 sill over the rough flooring. This is a variation of the sill shown in part *E* of Figure 32.

2. The sill construction under the bay window is to include the same wood members as shown for the dining-room wall. The brick veneer is to be placed next to the studs without sheathing.

3. The wall section over the bay window shows how the 4 by 8 timber is to be placed.

4. A 2 by 10 ridge is indicated.

5. A collar beam is indicated, together with its height and spacing.

6. The cornice construction is indicated.

7. The auxiliary rafter ties are specified and directions for their use given.

8. The screed-coat symbol indicates that the crawl-space earth must be level.

9. The 4′ 0″ dimension shows that the surface of the crawl-space earth must be that far below the bottoms of the living-room joists.

10. A bridging symbol indicates that such is required between joists.

11. The footing under the 10-inch part of the foundation is to be 20 inches wide.

12. Rough floor is indicated.

The foregoing example serves to indicate the relationship of section views to elevation and plan views—that the section views show what cannot be included in the other views.

SYMBOLS

Now that the relationships have been explained, we can understand that *both* plan and section views often indicate information about the same structural parts of a structure. For example, plan views show horizontal cut surfaces of a foundation, whereas section views show its vertical cut surfaces. Thus, the foundation is represented by both.

With the foregoing facts in mind, we can also understand that if a particular material symbol represents concrete on plan views, the same symbol represents concrete on section views. In other words, if we could cut through an actual foundation both horizontally and vertically, the material would look the same at both cut

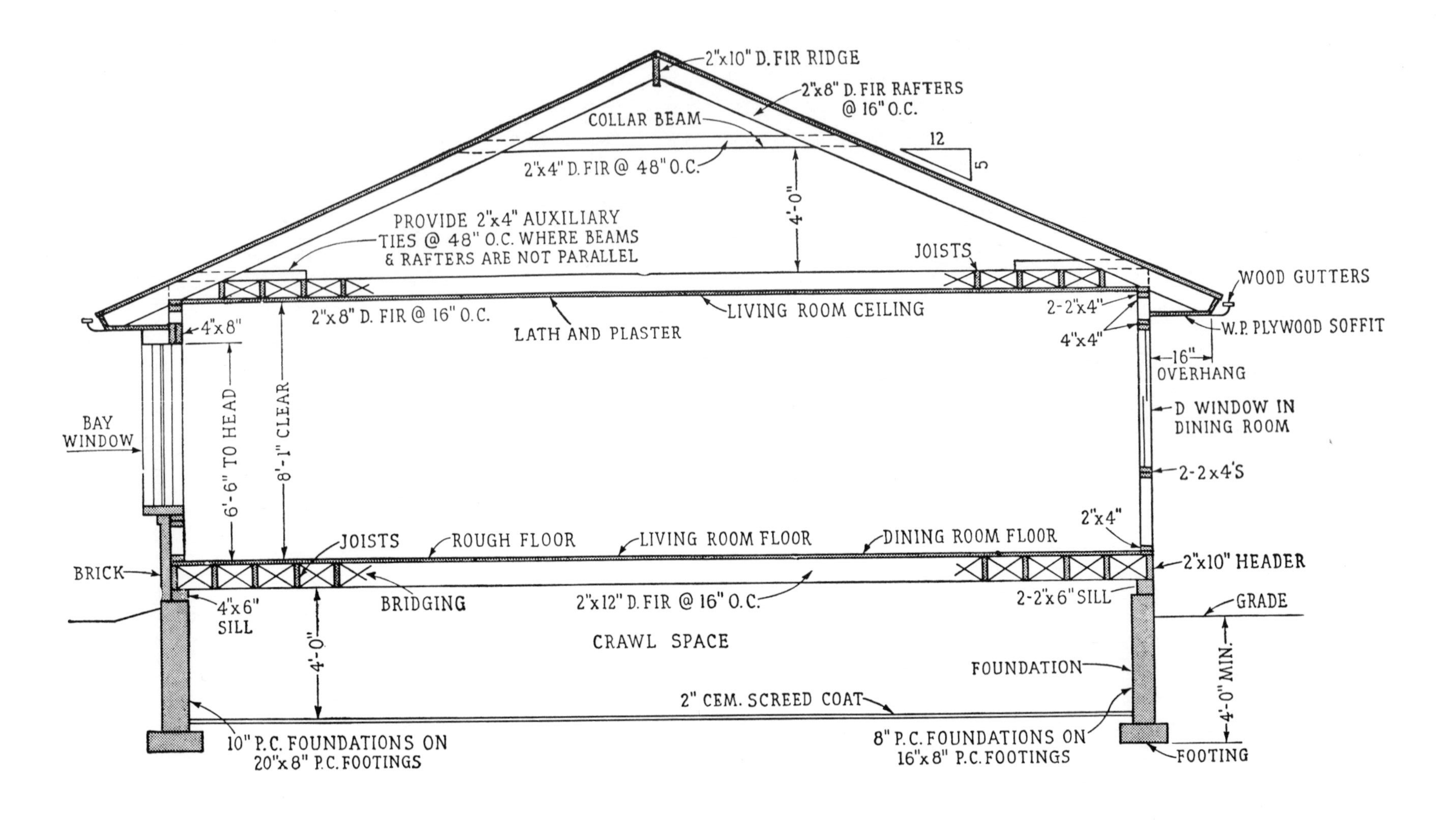
2"x10" D. FIR RIDGE
2"x8" D. FIR RAFTERS @ 16" O.C.
COLLAR BEAM
12
5
2"x4" D. FIR @ 48" O.C.
4'-0"
PROVIDE 2"x4" AUXILIARY TIES @ 48" O.C. WHERE BEAMS & RAFTERS ARE NOT PARALLEL
JOISTS
WOOD GUTTERS
2-2"x4"
W.P. PLYWOOD SOFFIT
4"x4"
16"
OVERHANG
4"x8"
2"x8" D. FIR @ 16" O.C.
LATH AND PLASTER
LIVING ROOM CEILING
BAY WINDOW
6'-6" TO HEAD
8'-1" CLEAR
D WINDOW IN DINING ROOM
2-2x4'S
2"x4"
JOISTS
ROUGH FLOOR
LIVING ROOM FLOOR
DINING ROOM FLOOR
2"x10" HEADER
BRICK
BRIDGING
4"x6" SILL
2"x12" D. FIR @ 16" O.C.
2-2"x6" SILL
GRADE
4'-0"
CRAWL SPACE
FOUNDATION
4'-0" MIN.
2" CEM. SCREED COAT
10" P.C. FOUNDATIONS ON 20"x8" P.C. FOOTINGS
8" P.C. FOUNDATIONS ON 16"x8" P.C. FOOTINGS
FOOTING
SECTION A-A
1/4" = 1'-0"
0 1 2 3 4 5 6 7 8 9 10 11 12 13 14
REDUCED SCALE

Figure 33

surfaces. Thus, the symbol shown at *B* in Figure 19 represents concrete on both plan and section views.

In the following, several typical section-view items are discussed in terms of the symbols which represent them.

Foundation The symbols shown at *B*, *C*, *E*, *K*, *Q*, *R*, *T*, and *U* in Figure 19 are all used to represent foundation materials.

Exterior Walls With the exception of the symbols shown at *G*, *H*, *J*, *M*, and *W*, all the symbols in Figure 19 are exterior-wall symbols.

Earth The symbol shown at *H* in Figure 19 is used only on section views and represents earth. For example, the symbol is frequently found in connection with foundation and grade symbols.

Glass The symbols shown at *W* in Figure 19 and at *B* in Figure 20 are used to indicate structural glass and glass blocks.

Steel Steel beams and columns are always indicated by symbols which look like the cut surfaces of these items. For example, the U-shaped objects shown in the symbol at *G* in Figure 20 are the symbols for a channel beam or column. The symbol for a steel-beam looks like an I. For example, such a symbol is shown at *D* in Figure 32.

Wood The symbols shown at *F* in Figure 19 are used to indicate rough-framing members. Generally, the cross-line symbol is used in connection with large members, such as headers, and the wavy symbol is employed for smaller members.

Plaster The symbol shown at *L*, in Figure 20, is used to indicate plaster. It is also used to represent mortar joints, mortar beds, and putty.

Gravel When gravel fills under concrete floor slabs or around footing draintile are required, the symbol shown at *O* in Figure 20 is used.

Dashed Lines As explained in connection with Figure 32, dashed lines represent hidden edges.

Reinforced Concrete When steel rods are to be used in concrete, their symbol looks like heavy solid lines or large dots, depending on whether the section views show their lengths or cut surfaces.

Insulation The general types of insulation are indicated by the symbols shown at *D*, *E*, *F*, and *W* in Figure 20.

Detail Symbols Section views often include items for which cut surfaces are not shown. In such cases, the items are indicated as in elevation views.

LINES

There is considerable variation in the way different designers employ line weights on section views. However, the most general line pattern includes the use of the heaviest lines to outline the parts composing the cut surfaces, the next lightest lines for symbols, and next lightest for projection lines between cut surfaces, and the lightest lines for dimensions.

TERMS AND ABBREVIATIONS

For the most part, the terms and abbreviations we have already learned, in our study of Chapters 3 and 4, will be ample for section views. A few new ones are necessary but we can learn them as they are encountered.

TYPES OF SECTION VIEWS

Four common types of section views are generally used:

1. The section views, like those shown in Figure 32, where related members, such as rafters and joists, are included in order to have a more picturelike appearance.

2. The section views, such as shown at *F* in Figure 31, where only the items within the area of the cut surface are included.

3. The section views which are combined with exterior views in order that a more picture-like appearance may be produced.

4. The section views of items such as fireplace chimneys, where a combination of elevation and plan views are included to give a more picturelike appearance.

CONSTRUCTION-PLAN STUDIES

The following illustrations represent actual section views as architects prepare them and as builders use them. In this book, because of page-size limitations, some of the section views have to be shown reduced in size. We can study the drawings just as though they were regular size to learn what they intend to have built.

Figure 34. The following observations will be helpful in learning to read section views of typical exterior walls. Readers are urged to locate all of the symbols, as they are mentioned, and to make sure that each explanation is thoroughly understood.

***A*-Part Materials** The section-view symbols show what the various items are to be made of:

Item	Material	Item	Material
Footing	*concrete*	Lathing	*gypsum*
Foundation	*concrete*	Framing	*wood*
Interior finish	*plaster*	Roofing	*asphalt shingles*
Roof boards	*⅝ wood*	Gutter	*G.I.*
Ext. finish	*asphalt finish*	Waterproofing	*asphalt*
Sheathing	*⅞ wood*	Kind of wood	*D.F.*
Finish floor	*oak*	Subfloor	*pine*
Soffit	*wood*	Molding	*wood*
Baseboard	*wood*		
Cellar floor	*concrete*		

***A*-Part Information** The following explanations constitute a partial example, so far as reading the section view is concerned.

House The section view in *A* indicates that the structure is to have one floor above a basement. A sloped roof is indicated.

Footing and Foundation Both are to be cast-in-place concrete, as indicated in the foregoing materials list. The footing is to be 8 inches deep, 16 inches wide, and have a key to prevent foundation slippage. The foundation is to be 10 inches thick and extend 7′ 6″ above the footing. The cellar floor is to be 4 inches thick.

Sill A 4 by 6 sill is to be set on a mortar bed, flush with the exterior edge of the foundation. The joists and header are to be 2 by 10s. Rough and finished flooring is required. The studs are to be supported by a 2 by 4 sill placed on top of the rough flooring.

Wall The studs are to be 2 by 4s. Gypsum lath and plaster are required. The wood sheathing is to be ⅞ inch thick. Asphalt shingles are to be applied directly to the sheathing.

Plate The plate is to be composed of two 2 by 4s. The ceiling joists are to be 2 by 8s.

Roof The rafters are to be 2 by 10s and the roof boards must be ⅝ inch thick with no space between them. The asphalt shingles are to be nailed directly to the roof boards.

Cornice The overhang of the rafters is to be 2′ 3″. The under framing is to be constructed using 2 by 4s. The soffit is to be ½ inch thick. When the kind of lumber for this particular member is not shown in the section view, we assume that the specs contain the necessary information. Hanging G.I. gutters are required.

Ceiling Heights The main-floor ceiling height is to be 8′ 2″ and the cellar ceiling height 7′ 6″.

Waterproofing The exterior faces of the foundation are to be waterproofed with asphalt, from footing up to the sheathing. The footing is to have a slope such that water will run away from the crack between the foundation and footing.

***B*-Part Materials** The section-view symbols show what the various items are to be made of:

Item	Material	Item	Material
Cellar floor	*none*	Lathing	*gypsum*
Interior finish	*plaster*	Roof boards	*⅝ wood*
Window frames	*wood*	Floor surface	*asphalt tile*
Exterior finish	*T and G siding*	Roofing	*wood shingles*
Fascia	*wood*	Soffit	*plywood*
Footing	*concrete*	Foundation	*concrete block*
Framing	*wood*	Gutter	*copper*
Floor	*concrete*	Floor fill	*gravel*
Baseboard	*wood*		

***B*-Part Information** The following informa-

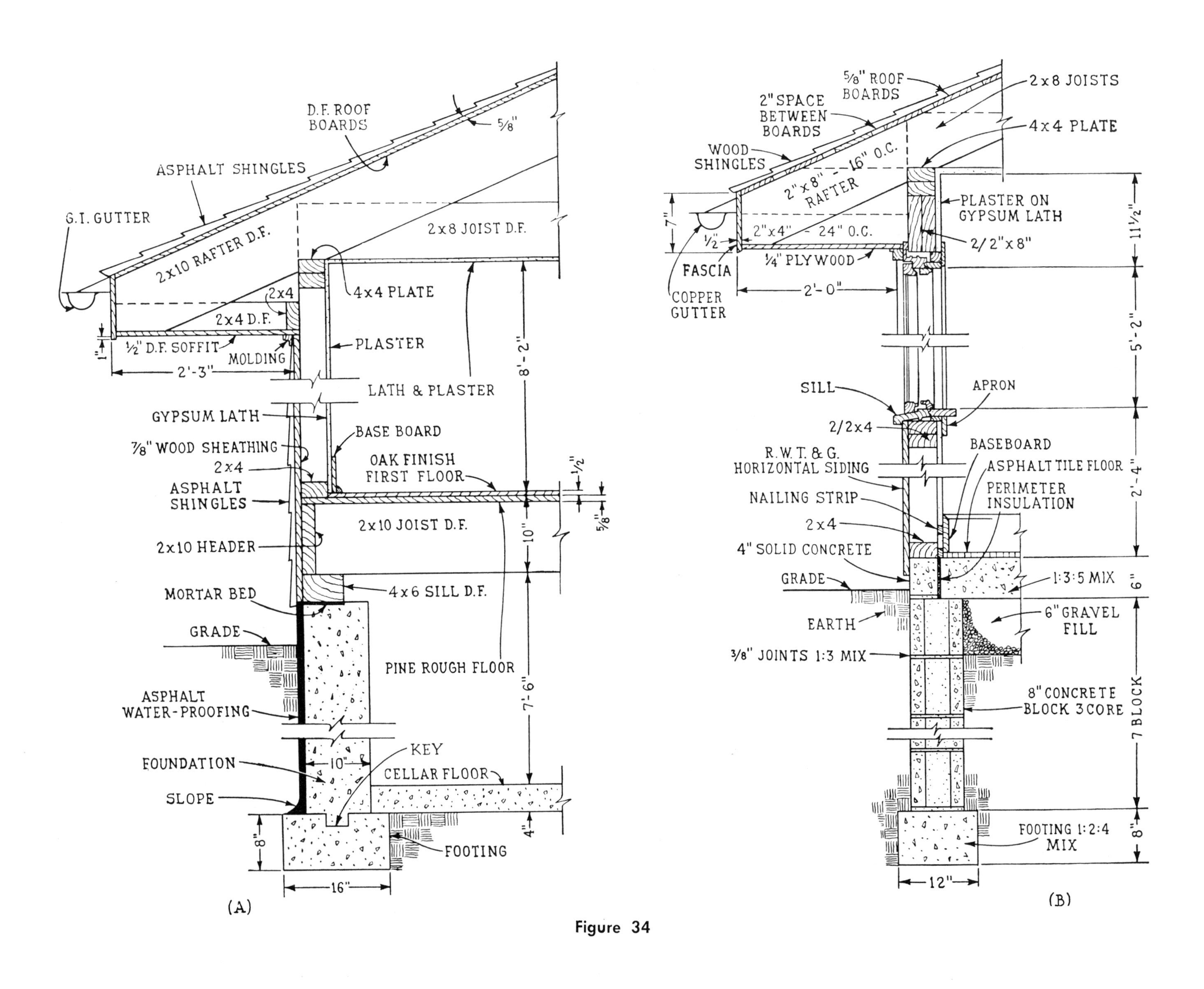
D.F. ROOF BOARDS
5/8"
ASPHALT SHINGLES
G.I. GUTTER
2x10 RAFTER D.F.
2x8 JOIST D.F.
2x4
2x4 D.F.
4x4 PLATE
1"
1/2" D.F. SOFFIT
MOLDING
PLASTER
2'-3"
8'-2"
LATH & PLASTER
GYPSUM LATH
BASE BOARD
7/8" WOOD SHEATHING
2x4
OAK FINISH FIRST FLOOR
1/2"
ASPHALT SHINGLES
5/8"
2x10 JOIST D.F.
2x10 HEADER
10"
MORTAR BED
4x6 SILL D.F.
GRADE
PINE ROUGH FLOOR
ASPHALT WATER-PROOFING
7'-6"
KEY
FOUNDATION
10"
CELLAR FLOOR
SLOPE
8"
4"
FOOTING
16"
(A)
5/8" ROOF BOARDS
2x8 JOISTS
2" SPACE BETWEEN BOARDS
4x4 PLATE
WOOD SHINGLES
2"x8" - 16" O.C. RAFTER
7"
PLASTER ON GYPSUM LATH
11 1/2"
2"x4" - 24" O.C.
1/2"
2/2"x8"
FASCIA
1/4" PLYWOOD
2'-0"
COPPER GUTTER
5'-2"
SILL
APRON
2/2x4
BASEBOARD
R.W.T. & G. HORIZONTAL SIDING
ASPHALT TILE FLOOR
2'-4"
NAILING STRIP
PERIMETER INSULATION
2x4
4" SOLID CONCRETE
GRADE
1:3:5 MIX
6"
EARTH
6" GRAVEL FILL
3/8" JOINTS 1:3 MIX
8" CONCRETE BLOCK 3 CORE
7 BLOCK
FOOTING 1:2:4 MIX
8"
12"
(B)

Figure 34

tion constitutes a partial example, so far as reading the section view is concerned.

House The structure in *B* is to have one story without a basement. A sloped roof is indicated.

Footing The footing is to be made of concrete composed of one part cement, two parts sand, and four parts crushed rock. It must be 12 inches wide and 8 inches deep.

Foundation The foundation is to be built of 8-inch concrete block which has three cores. Its height must be equal to seven blocks laid with ⅜-inch mortar joints. The mortar is to be composed of one part cement and three parts sand. At the top of the foundation, a solid block of concrete is to be used as the header for the floor.

Floor The floor is to be a 1:3:5 concrete slab over a 6-inch gravel fill. The edge of the floor is to have a 4-inch bearing on the foundation. Asphalt tile is to be cemented to the surface of the floor. Baseboard coves are not specified.

Wall The studs are to be 2 by 4s. Gypsum lath and plaster are required for interior finish. Tongued-and-grooved redwood horizontal siding is to be used for the exterior finish. Two 2 by 4s are to be used in the framing below window openings.

Plate This member is to be composed of two 2 by 4s. The ceiling joists are to be 2 by 8s.

Roof The rafters are to be 2 by 8s and the roof boards must be ⅝ inch thick with a 2-inch space between them. Wood shingles are to be nailed directly to the roof boards.

Cornice The overhang of the rafters is to be 2′ 0″. The under framing is to be constructed using 2 by 4s. The soffit is to be made of ¼-inch plywood. The fascia is to be 7 inches high and ½ inch thick. Hanging copper gutters are required.

Ceiling Height The ceiling is to be 8′ 5½″ above the floor.

Windows The section view indicates that 5′ 2″ casement windows are necessary.

Nailing Strips Two nailing strips are to be provided for the baseboards and two for each window.

Figure 35. This illustration includes the section views of the foundation, the floor, the exterior wall, and the doors for the one-story house shown in the elevation and plan views in Figures 12, 27*A*, and 27*B* of Chapters 3 and 4.

The *A* part of Figure 35 shows a section view which includes the foundation, the floor, and the wall. These sections, *Y-Y* and *X-X*, coincide with like symbols shown in Figures 27*A* and 27*B*. The following observations will be helpful in learning to read section views which are directly connected with other views.

***A*-Part Materials** The symbols show what the various items are to be made of:

Footings	*none*	Fl. surface	*asphalt tile*
Foundation	*concrete*	Int. finish	*dry wall*
Floor fill	*gravel*	Sheathing	*⅝ wood*
Framing	*wood*	Roof bds.	*⅝ wood*
Baseboard	*wood*	Shingles	*wood*
Fascia	*wood*	Siding	*wood T.G. vertical*
Soffit	*wood*		
Insulation	*loose*	Foundation insulation	*2″ perimeter*
Floor	*reinforced concrete*		

House The section view in *A* indicates a one-story house without a basement. A sloping roof is also indicated.

Foundation The bottom of the foundation is to be 12 inches wide and the top 4 inches wide. The total height must be 3′ 8″.

Floor The floor is to be a reinforced-concrete slab. A 4-inch fill of gravel is to be under the slab. There is to be a moisture barrier between the concrete and the gravel and 2-inch perimeter insulation which must extend 2′ 0″ below the surface of the concrete floor. The top of the floor must be 8 inches above grade.

Walls The studs are to be 2 by 4s. The interior finish is to be ½-inch dry-wall boards. The wood sheathing must be ⅝ inch thick. The exterior finish is to be 1- by 10-inch V-notched T and G vertical boards. The spaces between

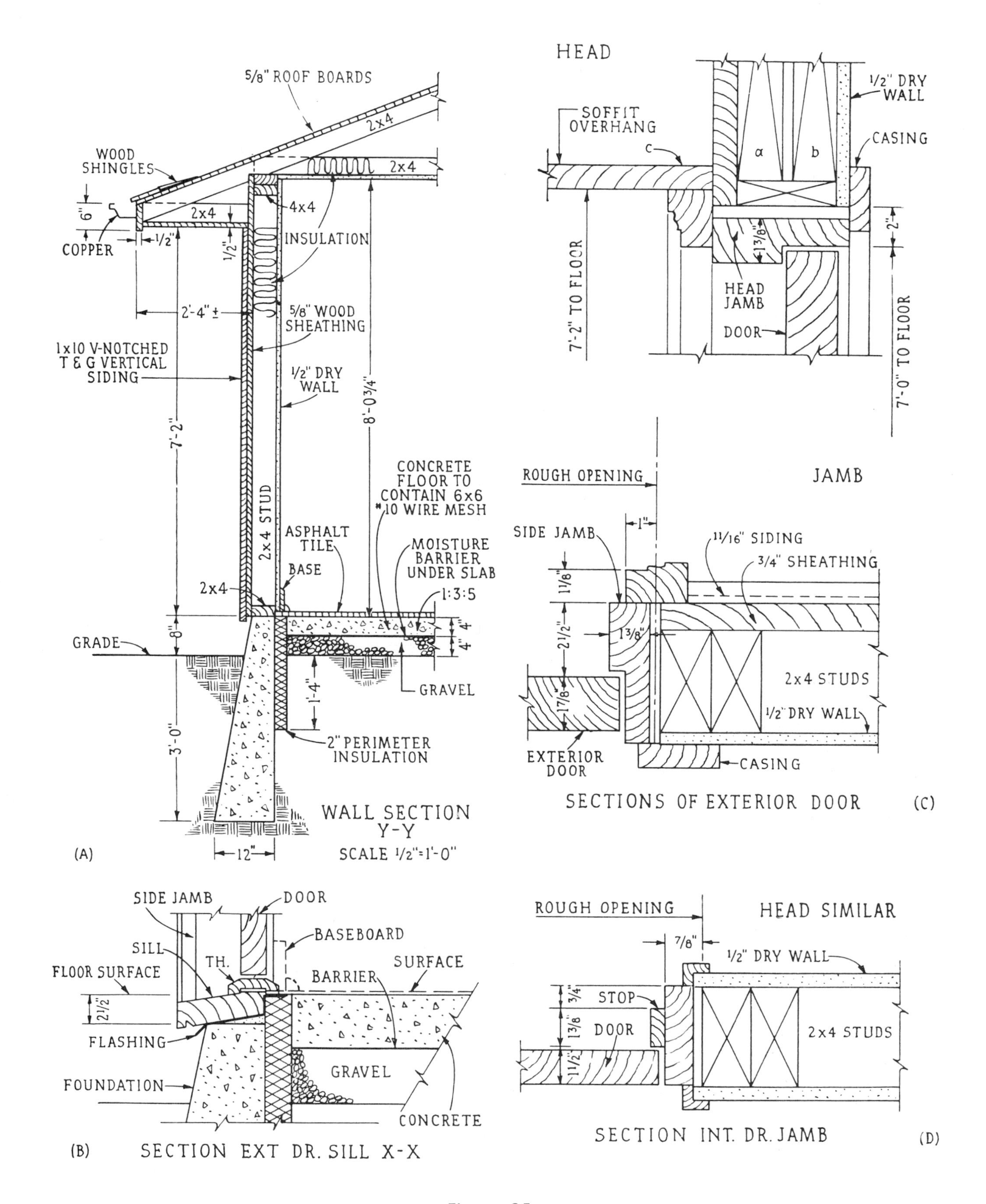

5/8" ROOF BOARDS
2x4
WOOD SHINGLES
2x4
6"
COPPER
1/2"
2x4
4x4
INSULATION
1/2"
2'-4" ±
5/8" WOOD SHEATHING
1x10 V-NOTCHED T & G VERTICAL SIDING
1/2" DRY WALL
8'-0 3/4"
7'-2"
CONCRETE FLOOR TO CONTAIN 6x6 #10 WIRE MESH
2x4 STUD
ASPHALT TILE
BASE
MOISTURE BARRIER UNDER SLAB
2x4
1:3:5
8"
4"
4"
GRADE
GRAVEL
1'-4"
3'-0"
2" PERIMETER INSULATION
WALL SECTION Y-Y
SCALE 1/2"=1'-0"
12"
(A)
HEAD
1/2" DRY WALL
SOFFIT OVERHANG
c
a
b
CASING
2"
13/8"
7'-2" TO FLOOR
HEAD JAMB
DOOR
7'-0" TO FLOOR
ROUGH OPENING
JAMB
1"
SIDE JAMB
11/16" SIDING
3/4" SHEATHING
1 1/8"
2 1/2"
1 3/8"
2x4 STUDS
1 7/8"
1/2" DRY WALL
EXTERIOR DOOR
CASING
SECTIONS OF EXTERIOR DOOR
(C)
SIDE JAMB
DOOR
BASEBOARD
SILL
TH.
FLOOR SURFACE
SURFACE
BARRIER
2 1/2
FLASHING
GRAVEL
FOUNDATION
CONCRETE
(B)
SECTION EXT DR. SILL X-X
ROUGH OPENING
HEAD SIMILAR
7/8"
1/2" DRY WALL
3/4"
STOP
1 3/8
DOOR
2x4 STUDS
1 1/2"
SECTION INT. DR. JAMB
(D)

Figure 35

studs are to be filled with fill-type insulation.

Plate The plate is to be made of two 2 by 4s. The ceiling joists are to be 2 by 4s.

Roof The rafters are to be 2 by 4s and the roof boards must be ⅝ inch thick with no space between them. Wood shingles are to be nailed directly to the roof boards.

Cornice The overhang of the rafters must be approximately 2′ 4″. This is indicated by the "plus or minus" sign after the 2′ 4″ dimension. The fascia and soffit are to be made of ½-inch wood. The depth of the fascia is to be 6 inches.

Dimensions Note that the plate height of 8′ 0¾″ and the soffit height of 7′ 2″ agree with like dimensions shown on the front elevation view in Figure 12.

From the foregoing observations, we can more clearly see the relationship among section, elevation, and plan views. The section views repeat many of the symbols shown on the other views, but they also include a great deal of information which could not be otherwise indicated. From now on in our study of section views, we shall pay more attention to structural information and less to symbols.

***B* Part of Figure 35.** This shows a section of the sill construction for the door marked 1 in the floor and foundation plans set forth in Figure 27*A*.

The section view of the sill was drawn to a larger scale than the wall section so that the depth of the required foundation notch could be shown and so that the method of constructing the sill could be indicated. The 2½-inch dimension means that in the notch the top of the foundation must be that far below the surface of the floor. The dotted symbol under the wood sill means that the sill must be set in a mortar bed. Note that flashing is required under and around the sill to keep water away from the perimeter insulation.

***C* Part of Figure 35.** This shows head and jamb section views for the same door discussed in connection with the *B* part of the illustration.

Head and jamb sections for doors and windows constitute the framing and finish materials at their tops and sides. For example, the head section in the *C* part of the illustration is visualized as though the top of door 1 in Figure 27 could be cut vertically and we could see the cut surface.

In the head section, *a* and *b* are the framing members over the door opening. We saw how framing is constructed over wall openings when we studied Figure 33. The member marked *c* is part of the cornice soffit (see Figure 12).

The ½-inch dry-wall specification means that a wallboard is to be used in place of lath and plaster. In the head section, the *top* of the door is indicated. We can more easily visualize this section if we locate the same parts in a similar and existing door. The 7′ 2″ dimension is the same as the like dimension shown in the *A* part of the illustration. The 7′ 0″ dimension indicates the height of the door.

In the jamb section, the two 2 by 4s are those on either side of the opening in the frame wall.

***D* Part of Figure 35.** This section shows a jamb section for interior doors. The head section, although not shown, is exactly the same. The rough-opening specification means that the distance shown is the opening between the 2 by 4s in the wall. Such 2 by 4s can be seen in the plan views shown in Figure 26.

Figure 36. This illustration, while not a standard type of section view such as is included with ordinary drawings, will help us to visualize the head, jamb, and sill sections of a typical window in a frame wall.

On the left-hand side of the illustration, elevation views of the exterior and interior of the window are shown. This combination of elevation views saves space and allows us to see what the window looks like from both sides.

The framing shown at the center of the illustration is called *isometric* because it was drawn so that we can see it as it would appear as part of the rough framing for a house. Note that the 2 by 4s are doubled on each side of the opening

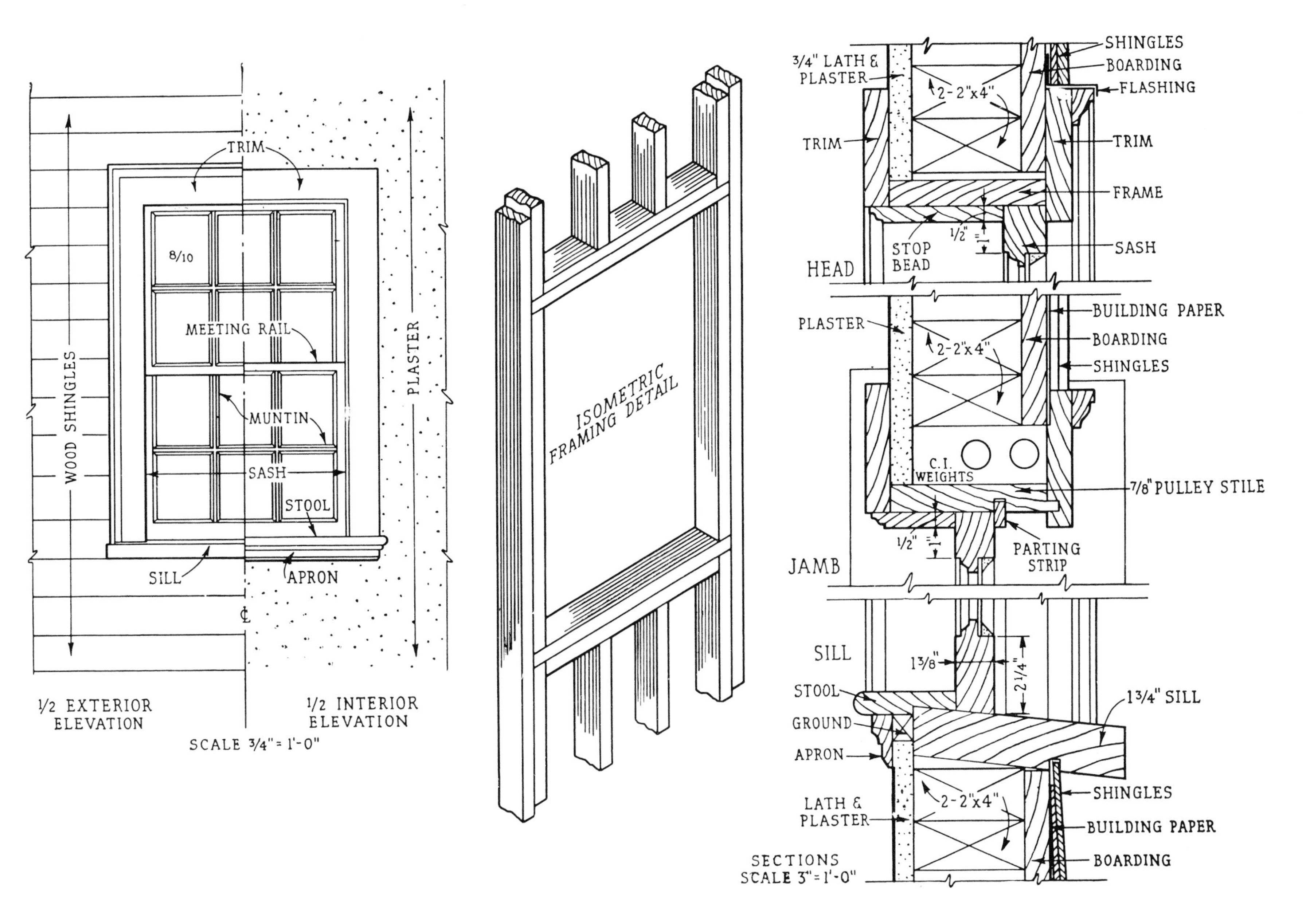
WOOD SHINGLES
TRIM
8/10
MEETING RAIL
MUNTIN
SASH
STOOL
SILL
APRON
PLASTER
1/2 EXTERIOR ELEVATION
1/2 INTERIOR ELEVATION
SCALE 3/4" = 1'-0"
ISOMETRIC FRAMING DETAIL
3/4" LATH & PLASTER
SHINGLES
BOARDING
FLASHING
2-2"x4"
TRIM
FRAME
1/2"
STOP BEAD
SASH
HEAD
PLASTER
BUILDING PAPER
BOARDING
SHINGLES
C.I. WEIGHTS
7/8" PULLEY STILE
PARTING STRIP
JAMB
SILL
1 3/8"
2 1/4"
STOOL
GROUND
APRON
1 3/4" SILL
LATH & PLASTER
SHINGLES
BUILDING PAPER
BOARDING
SECTIONS
SCALE 3" = 1'-0"

Figure 36

and that they are also doubled at the top and bottom of the opening. The vertical double 2 by 4s are indicated in Figure 26*A* and the horizontal double 2 by 4s in Figure 33.

In order to visualize the head section, shown at the right side of the illustration, we must imagine that the top of the window can be cut, and that we can see the exposed cut surface. The two horizontal 2 by 4s are those at the top of the isometric sketch. In order to visualize the jamb section, we must imagine that the side of the window can be cut so we can see the cut surfaces exposed. The same holds true for the sill.

The light vertical lines in the section views are to show how members in the head and sill coincide.

Readers are urged to learn the names of all members shown and to study them until their positions and relation to each other can be visualized. The sections will be easier to understand if readers will look at existing windows and locate such items as sill, aprons, parting strips, sash, etc.

While these section views are presented here to help us visualize window parts and understand how the windows fit into rough openings in walls, somewhat similar drawings are necessary in cases where designers specify windows which are not standard mill work—not available as stock items. Wood-working mills have hundreds of stock items to choose from, so special millwork is seldom necessary.

Figure 37. This illustration constitutes what can be called a *working-drawing example* of section views. It is so called because the information it shows is structural in nature. In order to understand the value of such a drawing, let us review Figures 26*A* and 26*B*.

On the sleeping and living level of the plan views, symbols indicate two flights of stairs. One flight is between the living and sleeping levels and the other between the living and utility levels. On the utility part of the plan views, near the symbol for the storage room, several other symbols are shown which concern the construction of the two flights of stairs. Note that a 6 by 6 post, doubled 2 by 10 header, and tripled 2 by 10 header are indicated. These structural indications are for the stairs. However, they are not complete enough to show carpenters how the structural work is to be done. The section view in Figure 37 supplies the necessary added information.

In order to visualize this section view, we must imagine that the stairs between the living and bedroom levels can be cut through along a line parallel to the arrow in the symbol. We must also imagine that we can look at the cut surface by standing in the hall and facing toward bedroom 1.

In Figure 37, we can locate several of the structural items indicated in the plan view. For example, we can see the 6 by 6 post, doubled 2 by 10 header, and tripled 2 by 10 header. In the plan view, both of these headers are shown at right angles to the stairs. Thus, we see their cut surfaces. Also, note the position of the space for door *L*, the linen closet partition, the sleeping-level hall, and the rear wall of the house.

The section view shows that 2 by 6s are required under the platform, between the foundation and the 6 by 6 post. There is to be a 1 by 6 sill under the 2 by 6s where they bear on the foundation. There is to be a 2 by 6 sill on top of the 6 by 6 post. The header must be supported by that sill. The carriages, parts which are to carry the treads and risers, are to be made from 2 by 10s. The ends of the carriages connect to the headers as shown.

Note that such dimensions as the 4′ 0″, the 3′ 8″, and the 4′ 4″ all check with like dimensions shown on the elevation and plan views. The $9' 7\frac{3}{16}''$ and $8' 0\frac{3}{4}''$ dimensions are shown to help carpenters lay out and check their work.

Figure 38. This illustration shows a section view of the stairs between the living and utility levels shown in Figure 26*A*. In order to visual-

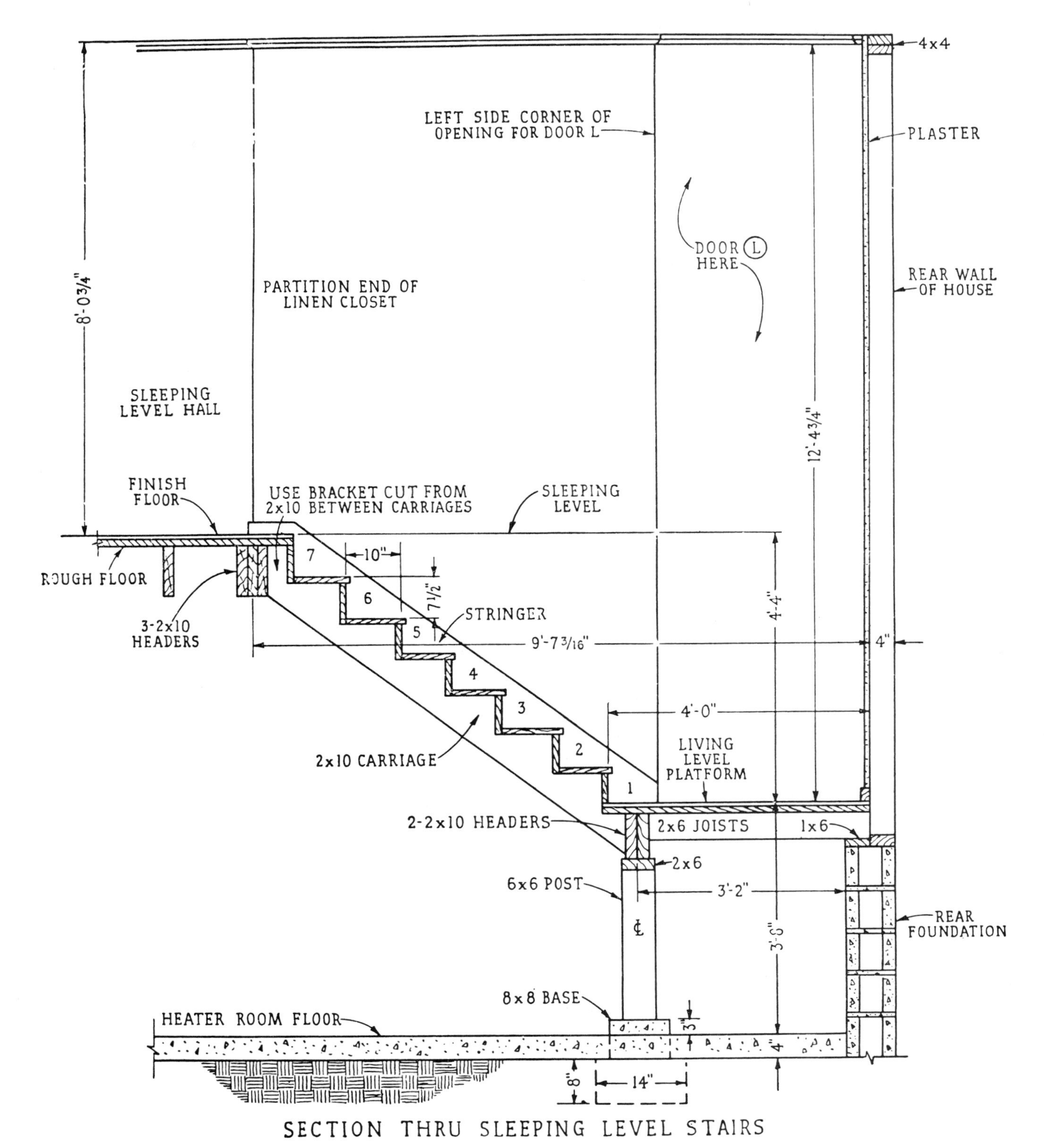

SECTION THRU SLEEPING LEVEL STAIRS

Figure 37

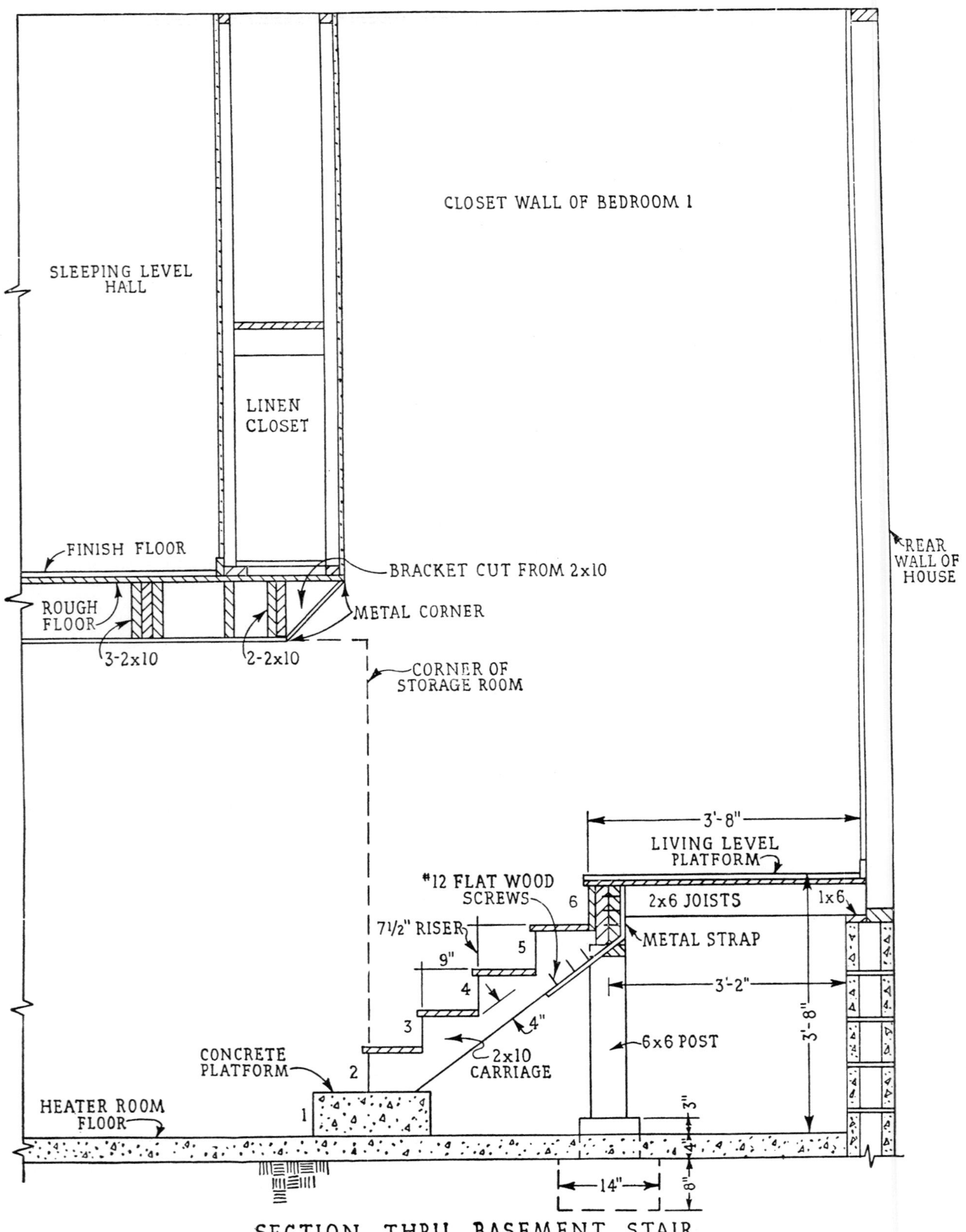

SECTION THRU BASEMENT STAIR

Figure 38

ize this drawing, we must imagine that the stairs can be cut through along a line parallel to the symbol arrow shown in Figure 26*A* and that we can look at the cut surfaces by standing in the living-level hall and looking in the direction of bedroom 1.

In Figure 38, we can locate some of the items indicated on the plan view and shown in Figures 26*A* and 26*B*. For example, we can see the 6 by 6 post, the rear wall of the house, the platform whose width is 3′ 8″, the sleeping-level hall, and the linen closet.

The section view shows that the platform joists are to be constructed of 2 by 6s which must be supported by the foundation at one end and by the 6 by 6 post at the other. The plan views do not indicate the concrete platform which forms one riser and one tread of the stairs. We can see that the carriages are to be fashioned from 2 by 10s and that they are to be fastened to the 4 by 10 (double 2 by 10) header with metal straps and No. 12 flat-headed wood screws. In order to have sufficient head room, it is necessary to bevel the ceiling corner under the linen closet. The section indicates that this construction must be accomplished by the use of brackets cut from 2 by 10s. Two 2 by 10s are required to support the sleeping-level floor above the stairs.

Readers are urged to study Figures 37 and 38 together with Figures 11, 26*A*, and 26*B* until they can visualize the section views easily. Once we learn such visualization, all section views will be easy to understand.

QUESTIONS AND ANSWERS

The questions in this chapter, unless otherwise noted, refer to Figure 39. However, it may be necessary to refer to Figures 14, 15, 29, and 30 in order to answer some of the questions.

As a means of making sure that you have learned to visualize section views and that you understand the relationship between them and elevation and plan views, answer each of the questions, either orally or in written form, and then check your reading with the *descriptive* answers shown.

Question 1 Suppose that we wanted to draw a double-headed arrow (such as shown in Figure 29 to indicate the section view in Figure 33) on the plan view in Figure 29, to indicate where the whole structure must be cut vertically through in order to show the cut surfaces in Figure 39. Where, and in what position, should the double-headed arrow be drawn?

Answer 1 The double-headed arrow must be drawn just above or below, and parallel to, the 19′ 4″ dimension in the living room and foyer. It would then indicate where to cut the structure in order to see the cut surfaces shown in the section view.

Question 2 What is the largest joist size indicated?

Answer 2 The largest joist size is 2 by 12 and is shown in the part of the section marked *F*.

Question 3 What does the symbol at *C* indicate?

Answer 3 The symbol indicates a partition which is to be framed using 2 by 4 studs, a 4 by 4 plate, and a 2 by 4 sill.

Question 4 What does the symbol at *G* indicate?

Answer 4 The symbol indicates that 2 by 4 collar beams must be used to stiffen the rafters.

Question 5 How many ridges are indicated and what size members are required for them?

Answer 5 Two ridges are indicated. The one shown at *a* must be a 2 by 10 and the one shown at *b* must be a 2 by 8.

Question 6 Suppose that we wanted to name the areas marked *A*, *B*, *D*, *E*, and *F* in the sectional view. What room names should we use?

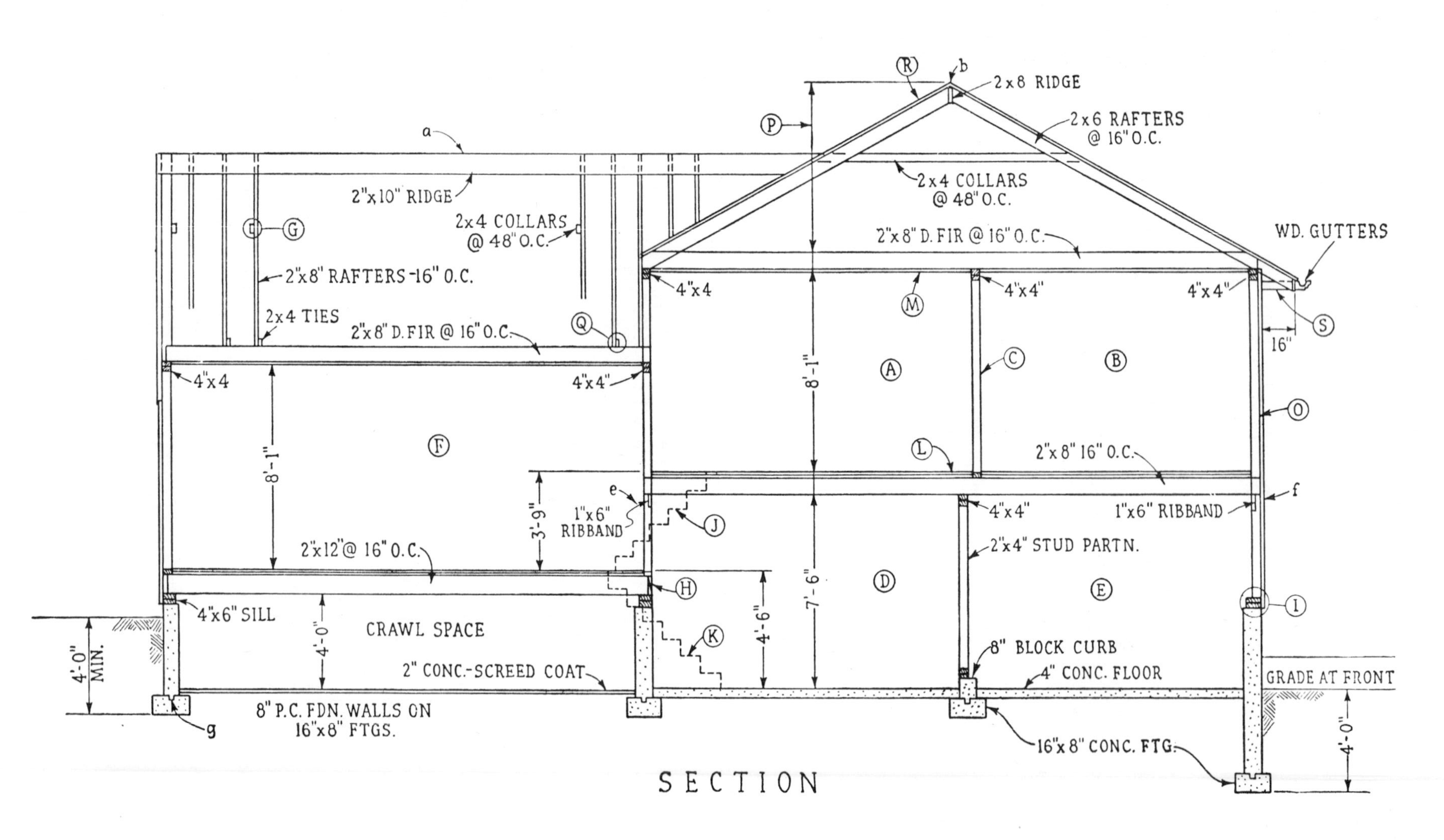
a
2"x10" RIDGE
2x4 COLLARS
@ 48" O.C.
G
2"x8" RAFTERS-16" O.C.
2x4 TIES
2"x8" D.FIR @ 16" O.C.
Q
h
4"x4
4"x4"
F
8'-1"
3'-9"
e
1"x6"
RIBBAND
2"x12"@ 16" O.C.
H
4"x6" SILL
4'-0"
CRAWL SPACE
2" CONC.-SCREED COAT
4'-0"
MIN.
g
8" P.C. FDN. WALLS ON
16"x8" FTGS.
R
b
2x8 RIDGE
P
2x6 RAFTERS
@ 16" O.C.
2x4 COLLARS
@ 48" O.C.
2"x8" D. FIR @ 16" O.C.
WD. GUTTERS
4"x4
M
4"x4"
4"x4"
S
16"
A
C
B
O
L
2"x8" 16" O.C.
f
8'-1"
4"x4"
1"x6" RIBBAND
J
2"x4" STUD PARTN.
D
E
7'-6"
4'-6"
I
K
8" BLOCK CURB
4" CONC. FLOOR
GRADE AT FRONT
16"x8" CONC. FTG.
4'-0"
SECTION

Figure 39

Answer 6 By studying Figure 29, we find that the imaginary cutting line goes through the living room, the foyer, and bedrooms 2 and 3. On Figure 30, the cutting line is in about the position of the 18′ 8″ dimension and goes through the crawl space, the den, and the garage. Thus, the following names should be used: *A*—Bedroom 3; *B*—Bedroom 2; *D*—Den; *E*—Garage; *F*—Living Room and Foyer.

Question 7 If we imagined that we could face the cutting surfaces, or cut surface, and that we could look through areas *A, B, D, E,* and *F,* what would we see?
Answer 7 From area *A*, we would see the partition between bedroom 3 and the hall. We would also see door 4. From area *B*, we would see the partition between bedrooms 1 and 2. From area *D*, we would see the partition between the den and the laundry. We would also see door 7. From area *F*, we would see the dining room, the partition between the foyer and the kitchen, door 4, and the first two risers of the stairs between the foyer and the bedroom-floor level.

Question 8 What does the small rectangular symbol at *Q* indicate?
Answer 8 It indicates a 2 by 4 rafter tie, and also that the rafters must be tied to prevent spreading.

Question 9 Note the stair indication at *K*. In what rooms are the top and bottom of these stairs to be?
Answer 9 The top is to be in the kitchen and the bottom in the den. This information is shown in Figures 29 and 30.

Question 10 Note the stairs indicated at *J*. In what rooms are the top and bottom of these stairs to be?
Answer 10 The top is to be in the hall at the bedroom floor level and the bottom in the foyer. This information is shown in Figure 29.

Question 11 What do the two close-together parallel lines at *R* indicate?
Answer 11 The lines indicate roof boards and shingles.

Question 12 What do the two close-together parallel lines at *O* indicate?
Answer 12 The lines indicate sheathing and exterior finish.

Question 13 What size header is required at *H*?
Answer 13 The header must be a 2 by 12 because it is to be used with 2 by 12 joists.

Question 14 What provision is to be made to prevent the foundation from slipping on the footing?
Answer 14 The footings must have keys in them, as indicated at *g*.

Question 15 What governs the distance between the top of the ridge and the tops of the ceiling joists on the bedroom-floor level?
Answer 15 This distance, as shown at *P*, is automatically established, because, as shown in Figure 14, the roof must slope at the rate of 6 feet in every 12 feet of run.

Question 16 What material is required at *S*?
Answer 16 As specified in Figure 14, the soffit is to be made of waterproof plywood.

Question 17 What do the two rectangular symbols at *I* indicate?
Answer 17 They indicate a 4 by 6 sill.

Question 18 What do the close-together parallel lines at *M* indicate?
Answer 18 The lines indicate lath and plaster.

Question 19 Where are ribbands required?
Answer 19 Ribbands are required in the wall between the foyer and bedroom 3, and in the exterior wall of bedrooms 1 and 2. On the section view, the two ribband indications are marked *e* and *f*.

Question 20 How far above the living-room floor is the bedroom floor level to be?
Answer 20 The dimension in the section view indicates that the bedroom-floor level is to be 3′ 9″ above the living-room floor.

CHAPTER SIX

Detail Views

In Chapter 5, section views were illustrated and explained as a means of showing the inside of various structural parts, so that carpenters and other mechanics can see exactly what is required. Detail views, on the other hand, make use of section views (also elevation and plan views) not only to show what the inside of various structural parts is like, but also to indicate where these parts are to be placed, the complete assembly of them, and their relationship to other parts. There are many instances where any difference between section and detail views is difficult to point out. There are other instances where detail views include more than assembly instructions. However, as implied in Chapter 5, once we understand section views we can easily visualize detail views. This fact will become more apparent as we study the following illustrations and explanations.

Just as there are many good and accepted ways of constructing frame walls, so far as their general type of construction is concerned, there are also good and accepted ways of assembly within the general types of construction. For example, the frame wall shown in Figure 32 is known as *modern braced framing*. In that particular type of framing, all of the section views shown at *B*, *C*, *D*, *E*, and *F*, demonstrate the general assembly practice. However, if particular kinds of exterior or interior finishes are necessary, some modification or special assembly instructions may have to be provided, so far as the placing of studs or other members are concerned. Detail views, as will be explained a little later in this chapter, are the means by which modifications or special assemblies are indicated.

Detail views, like elevation, plan, and section views, are also picturelike representations in which symbols, terms, and abbreviations are used as a means of indicating a great deal of information which cannot be included in any of the other views. Once we learn to read and visualize detail views, we will have progressed to the point where we can read any drawing on which general construction is indicated.

The purpose of this chapter is to show why detail views are necessary, how to visualize them, and how to read or use them with elevation, plan, and section views. We shall be concerned with the fundamentals of detail views rather than with their use in connection with the construction of popular types of structures. In other words, our aim is to learn what detail views are and their purpose. Once we understand such fundamental concepts, the applications will be clear.

WHY ARE DETAIL VIEWS NECESSARY?

The foregoing discussion contains some indication of why detail views are necessary. We can sum up all reasons, somewhat as we did in connection with section views, by saying that none of the other views, thus far studied, can be drawn to include the information detail views supply. The following examples will help us to understand why detail views are necessary:

Closet Framing The floor plans shown in Figure 26*A* indicate, for example, that a closet in bedroom 1 and a linen closet in the sleeping-level hall are required. (NOTE: The linen closet is indicated in the section views shown in Figures 37 and 38.) Because these two closets are to be close to each other, and because the space for both of them is complicated, detail drawings are necessary in order to show exactly how the framing must be erected. There are many possible ways in which the partitions could be framed or constructed. The purpose of the detail views is to show the particular or required way which will work to the best advantage with other parts of the general framing. There is a note near the closet symbol in Figure 26*A* which refers to detail views.

The plan view in Figure 26*A* also indicates coat and broom closets in the kitchen. There is to be a sliding door between the two closets. A note refers to detail views. The floor plan shows little about the two closets except their locations and their approximate sizes. Thus, detail views are necessary.

Kitchen Equipment The kitchen symbol shown in Figure 26*A* does not include equipment symbols. In many instances, especially where medium-sized kitchens are to contain a considerable amount of equipment, designers prefer to indicate such equipment by detail views. The note in Figure 26*A* indicates that all equipment information is shown on the detail views.

Roofs Section views, such as shown in Figures 33 and 39, can be drawn so that they indicate most of the required information about ordinary rafter-type roofs. For example, these two section views show the size of rafters, what type and size of ridges are required, how the rafters are to be tied, and how collar beams are to be used. One or two other section views could be drawn to indicate all other necessary information. However, where roof trusses are to be used, as indicated in Figure 12, section views cannot show all the required information in connection with the spacing and construction of the trusses. Thus, detail views are necessary.

Floors Where ordinary floors all on one level are concerned, the required information can be shown by the plan view symbols and by one or two section views. Where split-level houses are concerned, the floor framing is often complicated and cannot be shown on either plan or section views. In such cases, detail views are necessary.

HOW TO VISUALIZE DETAIL VIEWS

When we learned how to visualize elevation views, we imagined that we could stand directly in front of one side of a structure and look squarely at it. What we saw constituted an elevation view. When we learned how to visualize plan and section views, we imagined that a structure could be cut either horizontally or vertically through, and one part moved away so we could look at the cut surfaces. What we saw constituted plan or section views. In order to visualize detail views, we must learn two more and somewhat different ways of looking at various parts of a structure.

In general, and for the purposes of this book, detail views can be divided into two classes: the *placement* class, and the *assembly* class. These are explained in the following:

Placement Class This class of detail views indicates objects, such as kitchen equipment, in much the same manner as on elevation and plan views. The only difference is that detail views include elevation views of walls and plan views of floors which are only concerned with the placement of equipment. Thus, we imagine that we are looking squarely at walls of a room or directly down at the floors and that we can see the placement of equipment along the walls and on the floors.

Assembly Class On this class of detail views, the required information is generally indicated in elevationlike, planlike, or sectionlike drawings or by a combination of two or more of them. The following examples include the most commonly encountered ways by which assembly-class detail views are shown.

Elevationlike Views Any structural item, such as studs or built-in equipment, whose natural position is vertical, must be indicated in its proper position. Thus, the detail view looks something like the elevation view.

Elevationlike and Sectionlike Views Often, detail views include an elevation view to show the exterior appearance of an object or structural part and a section view to show the interior. In such cases, the two views are combined so that relationships can be seen.

Planlike and Sectionlike Views Use of this particular combination is required in a great many instances in order to show structural parts, such as wall framing, and how they are to be assembled.

Planlike Views When exact locations of joists and headers must be indicated, the detail views are drawn so that we must imagine we are looking down at them from directly above.

Planlike and Elevationlike Views Often, as in the case of roof trusses, a combination of plan and elevation views is necessary to indicate the required positions or spacings, as well as the method of building the trusses.

Visualization The visualization of detail views is not at all difficult because they are carefully made to give but one impression and because the representations are picturelike.

In order to visualize the assembly class of detail views, we must imagine certain circumstances.

The purpose of elevationlike views, such as shown in the *A* part of Figure 40, is to indicate positions, sizes, and spacings of various structural members. The views may be picturelike, as shown, or they may include only a flat view of the framing for one wall of a structure. In either case, we must imagine that all exterior and interior finish has been removed to reveal the skeleton or framework.

An elevationlike and sectionlike combination, as shown in the *B* part of Figure 40, must be visualized as we learned to visualize elevation and section views. By looking at the front and side elevations, we can easily visualize the backrest, seat, and drawer front of the object. We can understand that the detail view represents a seat which is to contain a large drawer and that the rear of the seat can be placed against a wall. In order to visualize the section view, we must imagine that parts of the backrest, seat, drawer, and bottom have been cut so that parts of them can be moved away to expose the cut surfaces *a, b,* and *c*. The cut surfaces show the interior construction of the seat. We can see how the two ends are to sup-

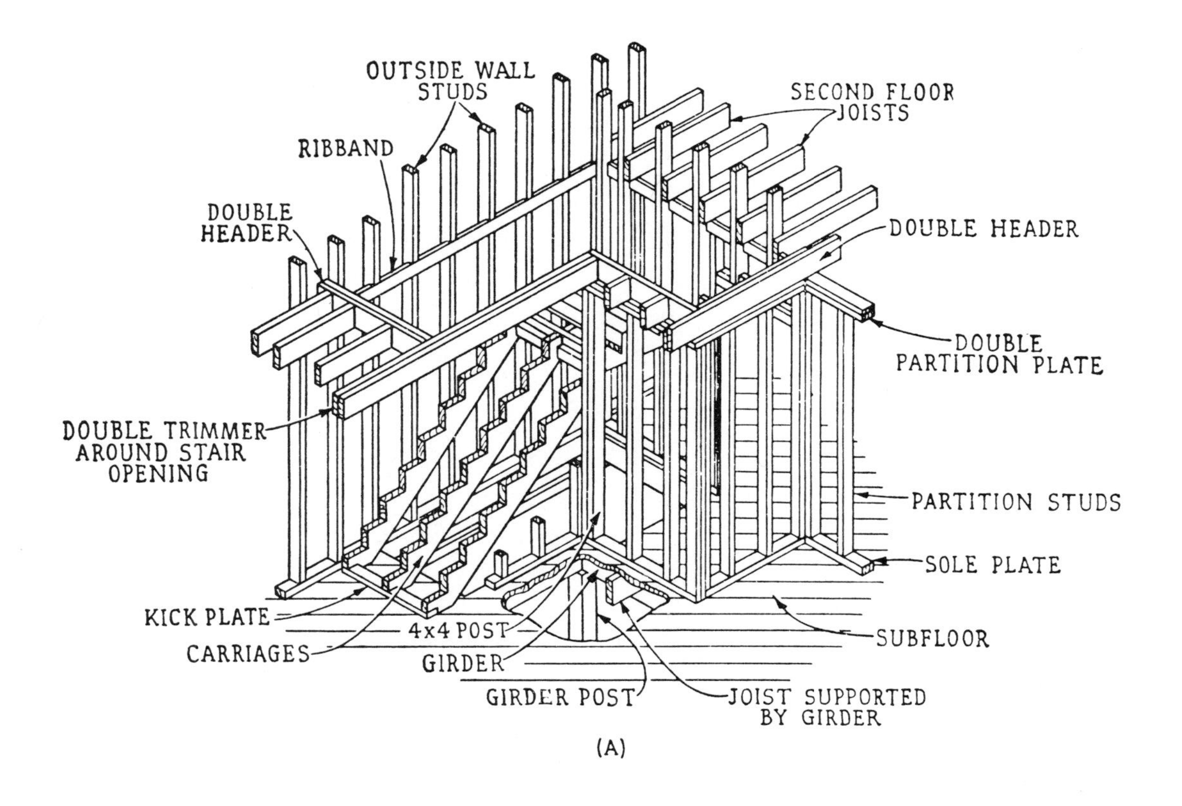
OUTSIDE WALL STUDS
SECOND FLOOR JOISTS
RIBBAND
DOUBLE HEADER
DOUBLE HEADER
DOUBLE PARTITION PLATE
DOUBLE TRIMMER AROUND STAIR OPENING
PARTITION STUDS
SOLE PLATE
KICK PLATE
4x4 POST
SUBFLOOR
CARRIAGES
GIRDER
GIRDER POST
JOIST SUPPORTED BY GIRDER

(A)

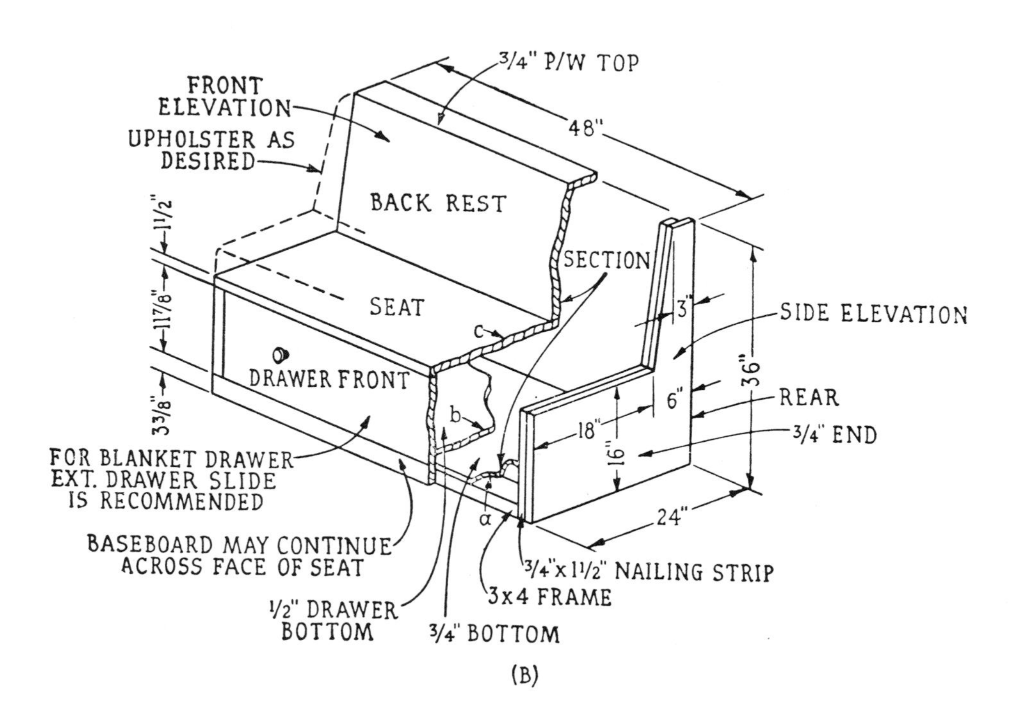
3/4" P/W TOP
FRONT ELEVATION
48"
UPHOLSTER AS DESIRED
BACK REST
1 1/2"
SECTION
SEAT
11 7/8"
3"
SIDE ELEVATION
c
DRAWER FRONT
36"
6"
REAR
3 3/8"
b
18"
3/4" END
16"
FOR BLANKET DRAWER EXT. DRAWER SLIDE IS RECOMMENDED
a
24"
BASEBOARD MAY CONTINUE ACROSS FACE OF SEAT
3/4"x 1 1/2" NAILING STRIP
3x4 FRAME
1/2" DRAWER BOTTOM
3/4" BOTTOM

(B)

Figure 40

port the backrest and seat and how the drawer is to be built. As in all elevation and section views, terms and dimensions are shown.

Planlike and sectionlike combinations show what we would see if we looked straight down at the *A* part of Figure 40 from a position directly above it. In this case, the cut surfaces of the studs appear as in section views, whereas the other members, which do not have cut surfaces, appear without a cut surface symbol in much the same manner as stairs appear in plan views.

Planlike views do not show any cut surfaces. For example, if studs were not included in the *A* part of Figure 40, we would not see any cut surfaces if we looked down at the framing from a point directly above it.

A planlike and elevationlike combination is visualized just as explained for such views. For example, if roof trusses are used in a structure, at least one plan view to show the spacing of the trusses and one elevation view to indicate the construction of the trusses are required.

RELATIONSHIP BETWEEN DETAIL VIEWS AND ELEVATION, PLAN, AND SECTION VIEWS

In the foregoing illustrations and explanations, we learned why and how some of the principles of elevation, plan, and section views are included in detail views. We also learned something of the relationship between detail and other views. In order to make that relationship more clear, we shall study some actual detail views.

First, however, let us review Figures 12, 27*A*, 27*B*, and 35 in connection with the one-story house.

Figures 12, 27*A*, and 27*B* indicate the positions and sizes of all windows and doors. However, the framing over these wall openings is not shown. Figure 35 indicates typical wall construction but does not show the framing over wall openings. Figures 27*A*, 27*B*, and 35 indicate that ½-inch plasterboard is required as interior finish, but do not show what size plasterboards to use, or how to apply such boards to avoid waste. Therefore, detail views must show the following information.

1. What sort of framing is required over the garage door, regular doors, and all windows.
2. How the framing over all openings is to be supported.
3. How the framing is to be constructed around window and door openings.
4. What size wallboard is to be used.
5. How the studs are to be spaced to avoid wallboard waste.

The *A* part of Figure 41 shows what is sometimes called a reflected ceiling plan. In this case, it is in connection with Figure 27*A*. In other words, the irregularly shaped area, which has its corners marked *a, b, c, d, e, f, g,* and *h,* is the same shape as the one-story house. The side *ah* contains the garage door. In Figure 12, we can see the elevation views of the sides marked *cb, ah,* and *gf*. These sides constitute the front elevation. This should help us to visualize the reflected ceiling plan.

Note that the reflected ceiling plan includes arrows and numbers; such as 1, 2, 3, 4, 5, 6, and 7, which point to the exterior sides; and numbers and arrows, such as 8, 9, 10, and 11, which point to some of the interior sides. In order to visualize the detail views explained in the following, it is necessary that we imagine we can look at the exterior and interior sides from the positions of the arrows.

Elevation 5 The *B* part of Figure 41 shows the required framing for the garage-door wall, *ah,* as seen from the position of arrow 5. The framing over the garage door is to be composed of two 2 by 12s which are to be supported by the number and grouping of studs indicated. The studs under the 2 by 12s are to be 6′ 9″ long, and those under the plate, 7′ 8¼″ long. Length of studs is found by subtracting plate thickness and header depth. Actual thicknesses should be used. Note that a 2 by 4 actually is

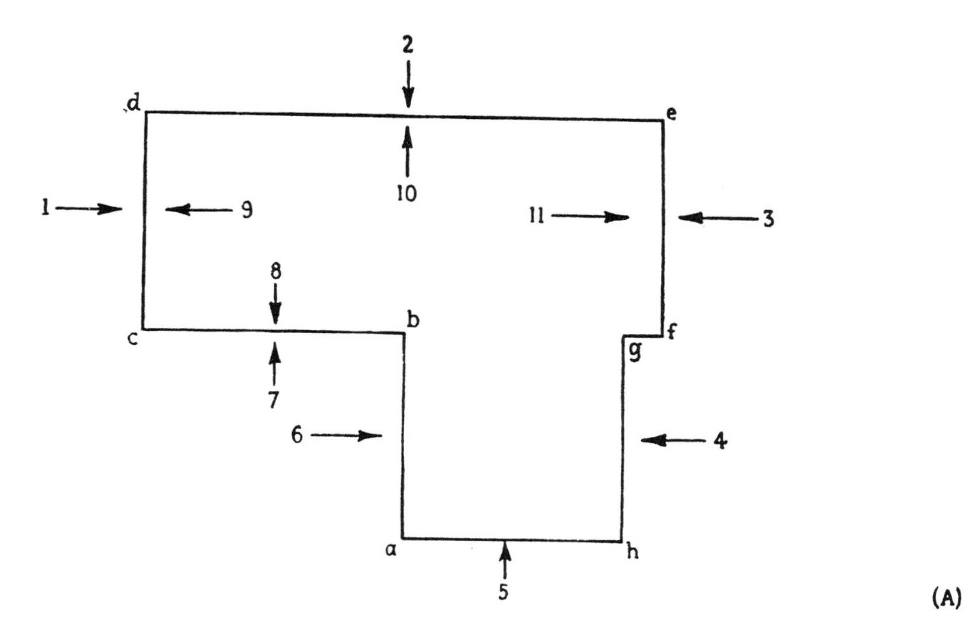

d
e
2
10
1
9
11
3
8
b
c
f
g
7
6
4
a
h
5
(A)

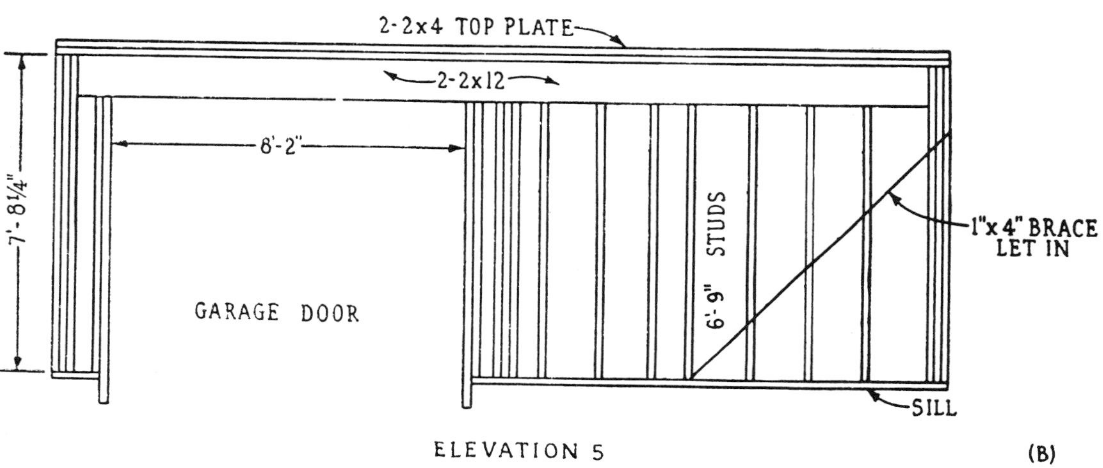

2-2x4 TOP PLATE
2-2x12
7'-8¼"
8'-2"
GARAGE DOOR
6'-9" STUDS
1"x4" BRACE LET IN
SILL
ELEVATION 5
(B)

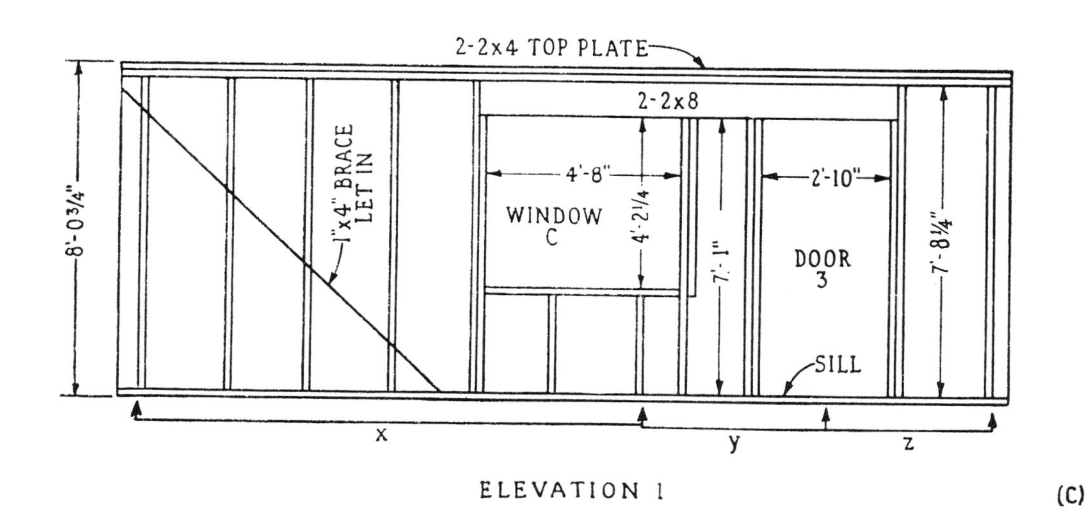

2-2x4 TOP PLATE
2-2x8
8'-0¾"
1"x4" BRACE LET IN
4'-8"
WINDOW C
4'-2¼
7'-1"
2'-10"
DOOR 3
7'-8¼"
SILL
x
y
z
ELEVATION 1
(C)

Figure 41

1½ by 3½; a 2 by 12, 1½ by 11¼; and a 2 by 8, 1½ by 7¼.

Elevation 1 This part of the illustration shows the required framing for the *dc* wall of the structure, as seen from the position of arrow 1. Figures 12 and 27*A* indicate that wall *dc* is to include door 3 and window *C*. The framing over the door and window openings is to be composed of two 2 by 8s which are to be supported by the number and grouping of studs indicated. Note that the studs under the 2 by 8s are to be 7′ 1″ long and that the distance from the bottom of the sill to the top of the plates is to be 8′ 0¾″. These dimensions check with like dimensions in Figures 12 and 35.

Elevation 4 The *A* part of Figure 42 shows the required framing for the *gh* wall of the structure as seen from the position of arrow 4. Over the window and door openings two 2 by 8s are required. They are to be supported by the number and grouping of studs indicated.

In like manner, detail views are used to show the framing for all other walls.

Elevation 9 The *B* part of the illustration shows the required interior finish for the *dc* wall of the structure as seen from the position of arrow 9. Note that this is the *interior* of the same wall shown in the *C* part of Figure 41. The interior finish is to be ½-inch gypsum board of the 4 by 8 and 4 by 12 foot size. Note the dimensions *x*, *y*, and *z* in the *C* part of Figure 41. Also note the like dimensions in the *B* part of Figure 42. These dimensions show that the stud locations as well as the length of wall *dc* were planned to avoid waste of the standard-size gypsum boards. It should be noted that the height of the walls was also planned to avoid wallboard waste.

Elevation 11 The *C* part of Figure 42 shows the interior finish for the *ef* wall of the structure as seen from the position of arrow 11.

The foregoing illustrations and explanations bring out the relationship between detail and all other views. As we now understand, detail views are simply an additional means of showing necessary information in construction plans.

SYMBOLS

For the most part, symbols on detail views are the same as those found in section views. The few differences are explained in the following:

Joist Symbols On planlike detail views, joists are indicated by two parallel lines. No other material symbol is included. We must imagine that we are looking straight down at them and that the parallel lines represent their edges.

Wall Symbols On planlike detail views, walls and foundations are sometimes shown by parallel lines without material symbols. The reason for this is that they are shown merely to provide the complete picture. For example, the ends of joists are generally shown supported by walls and foundations.

Cut Surfaces Planlike and sectionlike detail views, as previously explained, frequently make use of cut surfaces, as in section views. However, the section views are generally *horizontal* cuts which we must visualize as though we are looking straight down at them. For example, studs and other framing members which are to be parts of walls and partitions are shown as if the walls or partitions had been cut horizontally so we can see how the framing is to be assembled. Plaster, doors, trim, etc., are also shown as if cut horizontally. Plan-view material symbols are employed.

Equipment Symbols On planlike detail views, equipment is indicated by the same symbols as appear in regular plan views. For example, kitchen-equipment symbols consist of square or rectangular shapes. On elevationlike detail views, equipment is drawn as it would appear if we could see an actual installation along the walls of a kitchen.

Roof Framing On planlike and elevationlike detail views, roof rafters and trusses are some-

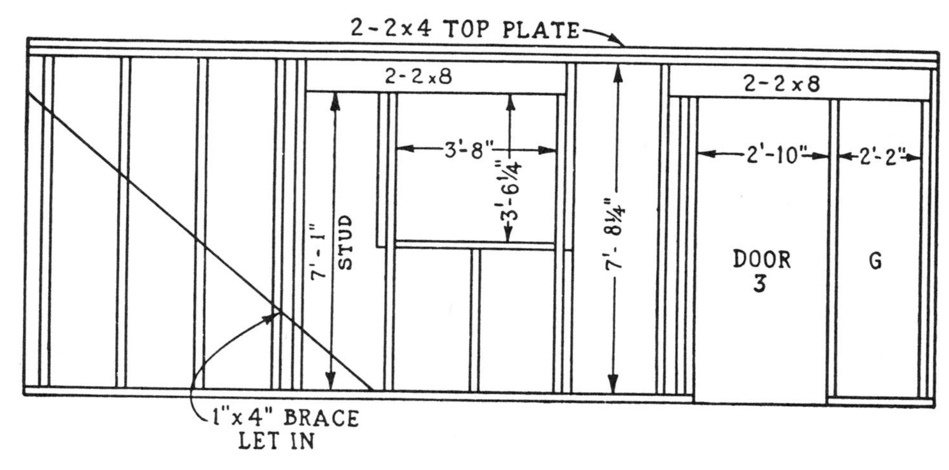

ELEVATION 4

(A)

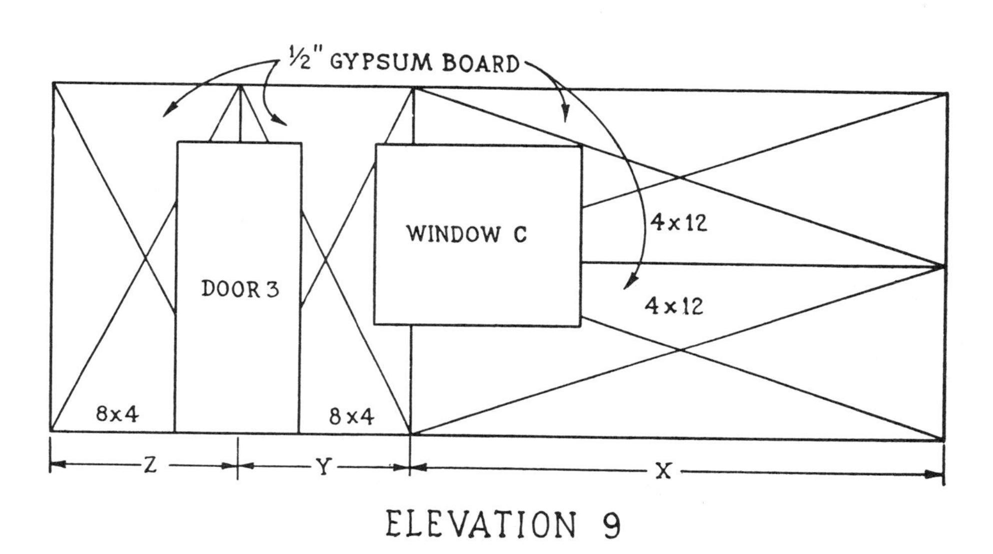

ELEVATION 9

(B)

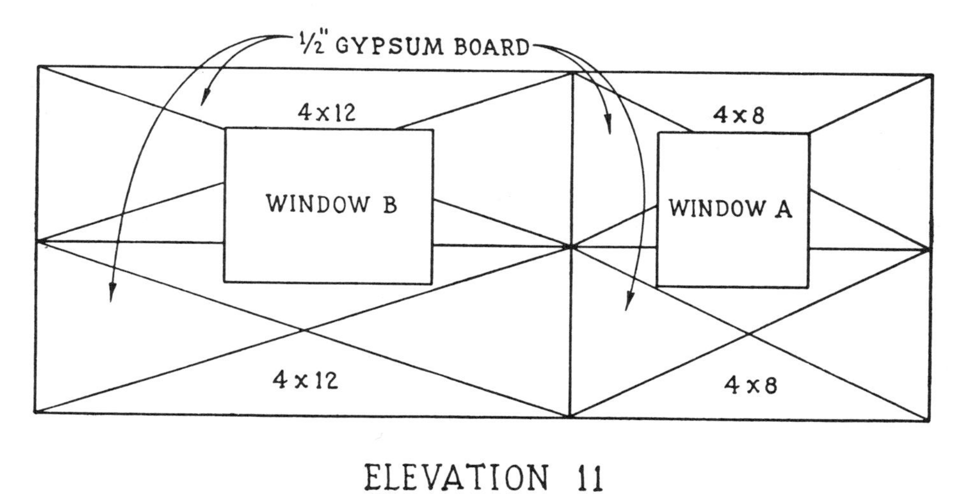

ELEVATION 11

(C)

Figure 42

times shown, as in regular plan views, except that no cutting is involved and single lines represent them. If structural information is indicated for trusses, the members are shown as though we could stand to one side of a truss and look squarely at it.

Floor Framing Planlike detail views are frequently drawn to show the required arrangement of joists, bridging, etc. In most cases, the members are indicated by single lines.

Wall Framing In our study of Figures 41 and 42, we learned that studs, sills, plates, and beams (such as the 2 by 12s and 2 by 8s used over wall openings) are shown in elevationlike detail views and that these members are indicated by parallel lines.

TERMS AND ABBREVIATIONS

We are already acquainted with most of the terms and abbreviations which appear on detail views. Where unfamiliar items appear, they are explained as they occur.

CONSTRUCTION-PLAN STUDIES

The following illustrations represent actual and typical detail views as architects prepare them and as builders use them. In this book, because of page size limitations, some of the detail views have to be shown reduced in size. We can study the drawings just as though they were regular size to learn what they intend to have built.

Figure 43 This illustration constitutes a placement-class detail view. Note that it is mostly concerned with the plan and elevation views of kitchen equipment and that only enough other symbols are indicated to provide proper relationship or surroundings for the equipment.

In Figure 26*A*, the general kitchen symbol does not include symbols for such equipment as the sink, the range, the refrigerator, cupboards, etc. This is a case where the designer preferred to use larger detail views in which to indicate all such equipment. As shown in Figure 26*A*, the kitchen is only 13′ 6″ by 8′ 0″. A room as small as that, while large enough for an efficient kitchen, makes careful planning necessary. Thus, the larger detail views better serve the purpose. A note refers to details.

The *A* part of Figure 43 shows the same general plan as in Figure 26*A*. Note the *H* door, the door to the dining space, and the coat and broom closets. The detail view is in the same position as in the regular plan view.

Note that all equipment is clearly indicated by labeled symbols. Each symbol scales accurately as per the dimensions shown. For example, the lengths of the sink, dishwasher, washer, and dryer total 13′ 0″. The overall dimension is 13′ 6″. That leaves 6 inches for space between equipment.

Along the range and refrigerator wall, two cabinets or cupboards, whichever we choose to call them, are indicated. The dashed lines mean that two cabinets are actually involved: one whose top is level with the top of the range, and one higher up on the wall. The dashed lines shown in connection with the dishwasher and dryer indicate that cabinets are required above them.

Note that the sink is under the window and that, from such a detail view, plumbers can plan the locations of hot- and cold-water pipes, drains, and gas pipes. Note, too, that the dimensions check with the dimensions shown in Figure 26*A*. There is a specification to indicate that the owner of the house is to select the linoleum.

The *B* part of Figure 43 shows an elevation view of the equipment for the range and refrigerator wall. We can see the upper cabinets which are indicated by dashed lines in the planlike detail view.

The range and cabinet tops are to be 3′ 0″

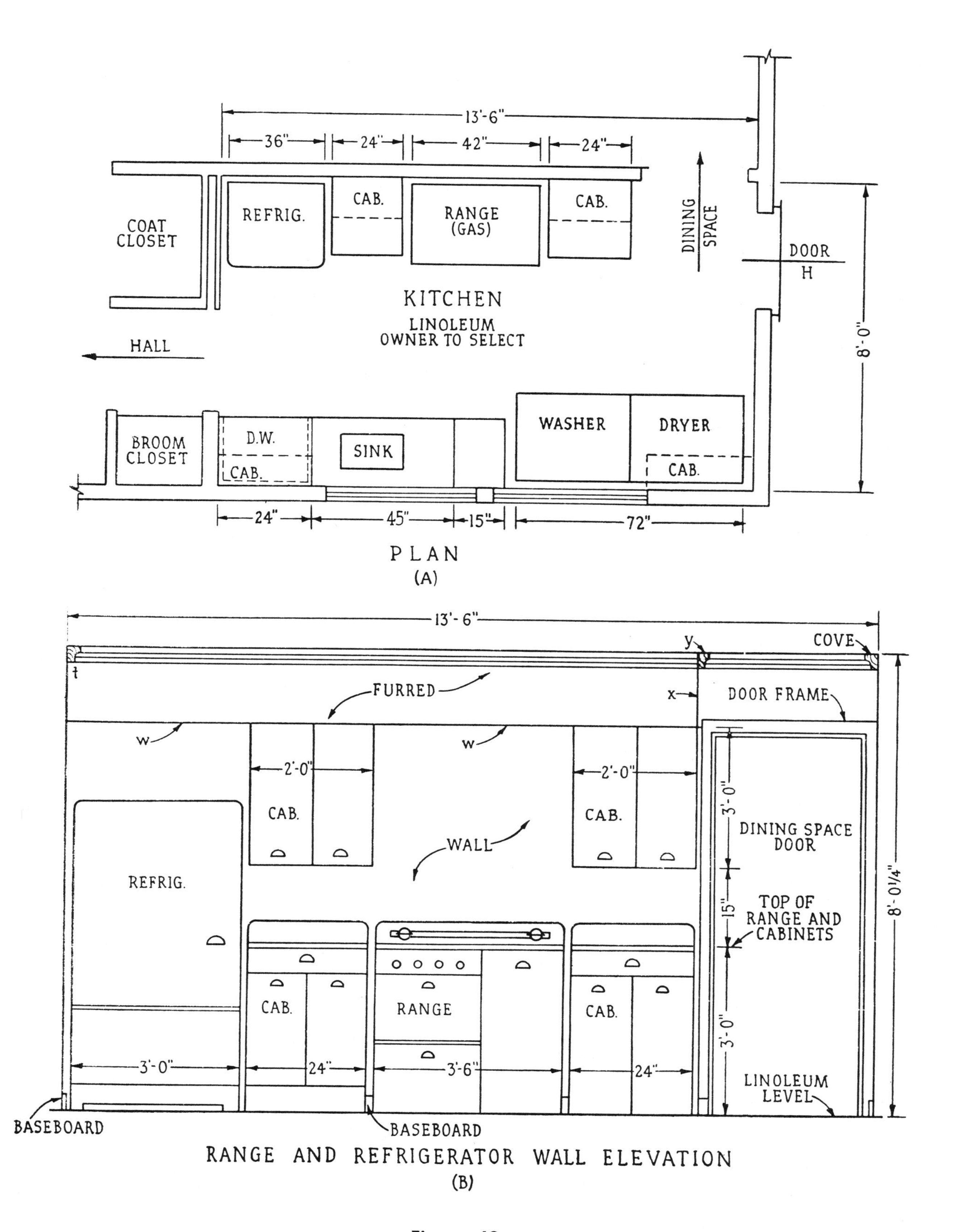
13'-6"
36"
24"
42"
24"
COAT CLOSET
REFRIG.
CAB.
RANGE (GAS)
CAB.
DINING SPACE
DOOR H
8'-0"
KITCHEN
LINOLEUM OWNER TO SELECT
HALL
BROOM CLOSET
D.W.
CAB.
SINK
WASHER
DRYER
CAB.
24"
45"
15"
72"
PLAN
(A)
13'-6"
COVE
t
y
FURRED
x
DOOR FRAME
w
w
2'-0"
2'-0"
CAB.
CAB.
WALL
DINING SPACE DOOR
3'-0"
15"
3'-0"
REFRIG.
TOP OF RANGE AND CABINETS
8'-0¼"
CAB.
RANGE
CAB.
3'-0"
24"
3'-6"
24"
LINOLEUM LEVEL
BASEBOARD
BASEBOARD
RANGE AND REFRIGERATOR WALL ELEVATION
(B)

Figure 43

above the floor. The bottoms of the upper cabinets are to be 1′ 3″ above the range top. These cabinets are to be 3′ 0″ high.

The "furred" notation means that above the upper cabinets, the wall from *x* to *t* is to be built out flush with the fronts of the cabinets. In other words, between the two upper cabinets and above the refrigerator, the ceiling height is to be as indicated at *w* but only for a distance equal to the depth of the cabinets. At *y*, a molding between the ceiling and the wall is indicated. The vertical line at *x* indicates that the furring ends at the left-hand side of the door. Between *x* and *t*, the molding is next to the real ceiling.

The kind of equipment, such as the manufacturer's name, and the materials to be used, are covered in the specs.

Figure 44 The *A* part of this illustration constitutes an assembly-class planlike detail view. Note that it is mostly concerned with the assembly of joists and headers and that the joists, headers, and foundations are all indicated by means of the symbols previously explained.

In Figures 37 and 38, we studied section views of the two flights of stairs indicated in Figure 26*A* for the split-level house. The section views show how the stairs are to be built, but they do not give much information relative to the floor framing or the headers. Therefore, detail-view information is necessary. This is a good example of the necessity for detail views.

First, let us study the relationship between the *A* part of Figure 44 and Figure 37. The three 2 by 10s, shown at *a* in Figure 44, are the same as the header shown at the top of the stairs in Figure 37. The header indicated by the long-and-short dashed line, in Figure 44, is the same as the header shown at the foot of the stairs in Figure 37.

Next, let us study the relationship between the *A* part of Figure 44 and Figure 38. The three 2 by 10s, shown at *a* in Figure 44, are the same as the header shown at the left side of the linen closet in Figure 38. The header indicated by the long-and-short dashed lines in Figure 44 is the same as the header shown at the top of the stairs in Figure 38.

The 4-inch post indicated in Figure 44 is to be supported by the doubled 2 by 10s header shown at top of the stairs in Figure 38.

The doubled 2 by 10s at *b* in Figure 44 are the same as the doubled 2 by 10s under the linen closet in Figure 38.

All headers (tripled 2 by 10s and doubled 2 by 10s) indicated in Figure 44 are at the bedroom level shown in Figure 26*A*.

The two 2 by 10s shown at *c* in Figure 44 are to be supported by the three 2 by 10s, at one end, and by the 4 by 4 post, at the other end. This header is shown between the two staircase openings. The two 2 by 10s, in turn, are to support the partition between the two staircases. This is indicated in Figure 26*B*.

The foundation indicated at *d* in Figure 44 is the foundation shown between the garage and furnace room in Figure 26*B*.

By studying Figures 26*A*, 26*B*, 37, 38, and 44, all the necessary information relative to the two staircases and the framing required for them can be visualized. It is important for readers to understand the relationships of all drawings and the structural work to be done.

The *B* part of Figure 44 shows regular section views of the type of trim to be used for the house represented in Figures 11, 26*A*, and 26*B*. Examples of the use of this trim are shown in Figure 43, and in the following illustrations. The backband trim is to be used around windows and doors. The corner trim is to be used at wall corners where plasterboards meet. The cove trim is to be used where the walls and ceiling meet. The base trim is to be used where the walls and floors meet. Note that the section views are shown full-scale or actual size.

Figure 45 This illustration constitutes an assembly-class plan- and sectionlike detail view.

In the sleeping-level part of Figure 26*A* two

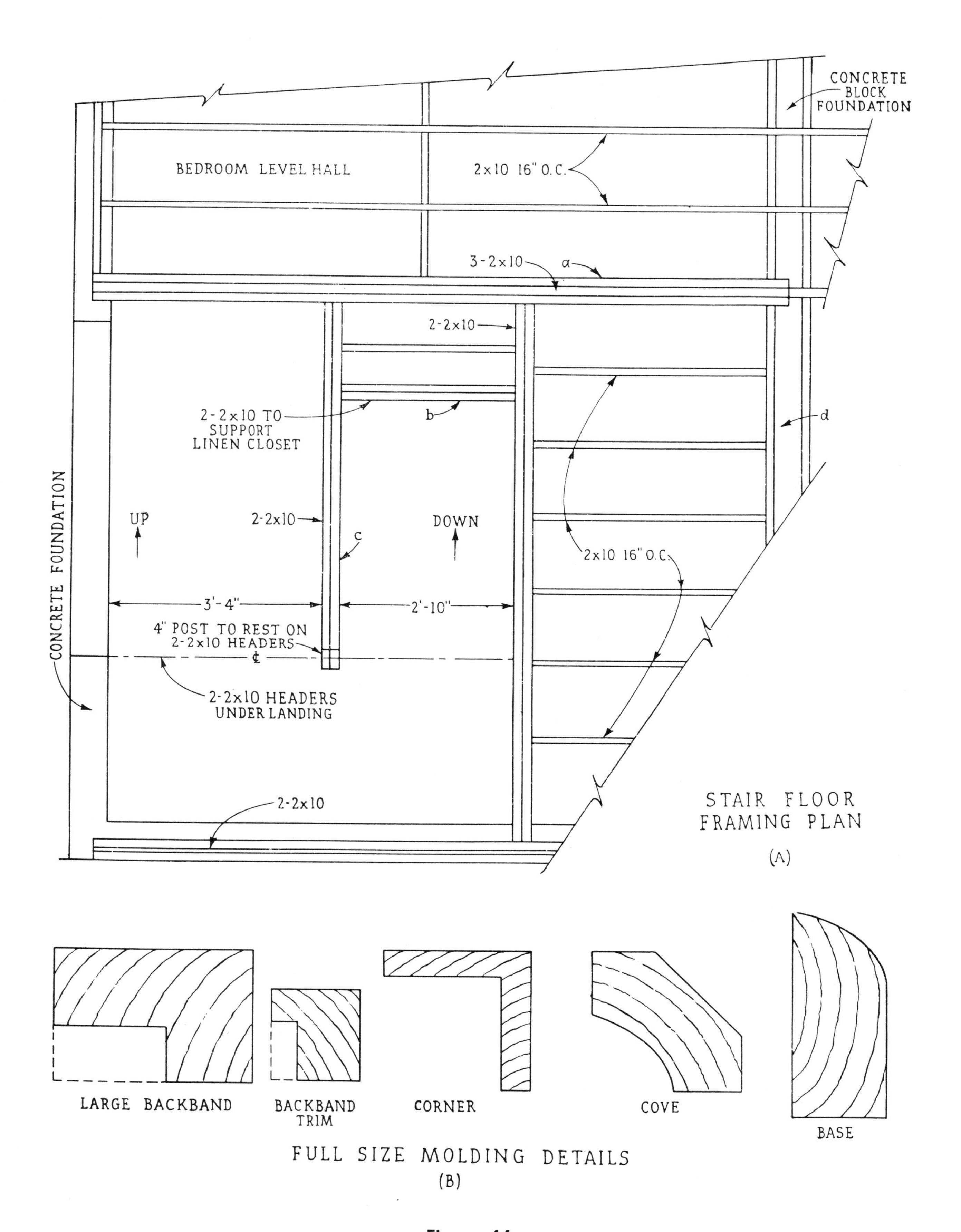
CONCRETE BLOCK FOUNDATION
BEDROOM LEVEL HALL
2x10 16" O.C.
3-2x10
a
2-2x10
2-2x10 TO SUPPORT LINEN CLOSET
b
d
CONCRETE FOUNDATION
UP
2-2x10
c
DOWN
2x10 16" O.C.
3'-4"
2'-10"
4" POST TO REST ON 2-2x10 HEADERS
℄
2-2x10 HEADERS UNDER LANDING
2-2x10
STAIR FLOOR FRAMING PLAN
(A)
LARGE BACKBAND
BACKBAND TRIM
CORNER
COVE
BASE
FULL SIZE MOLDING DETAILS
(B)

Figure 44

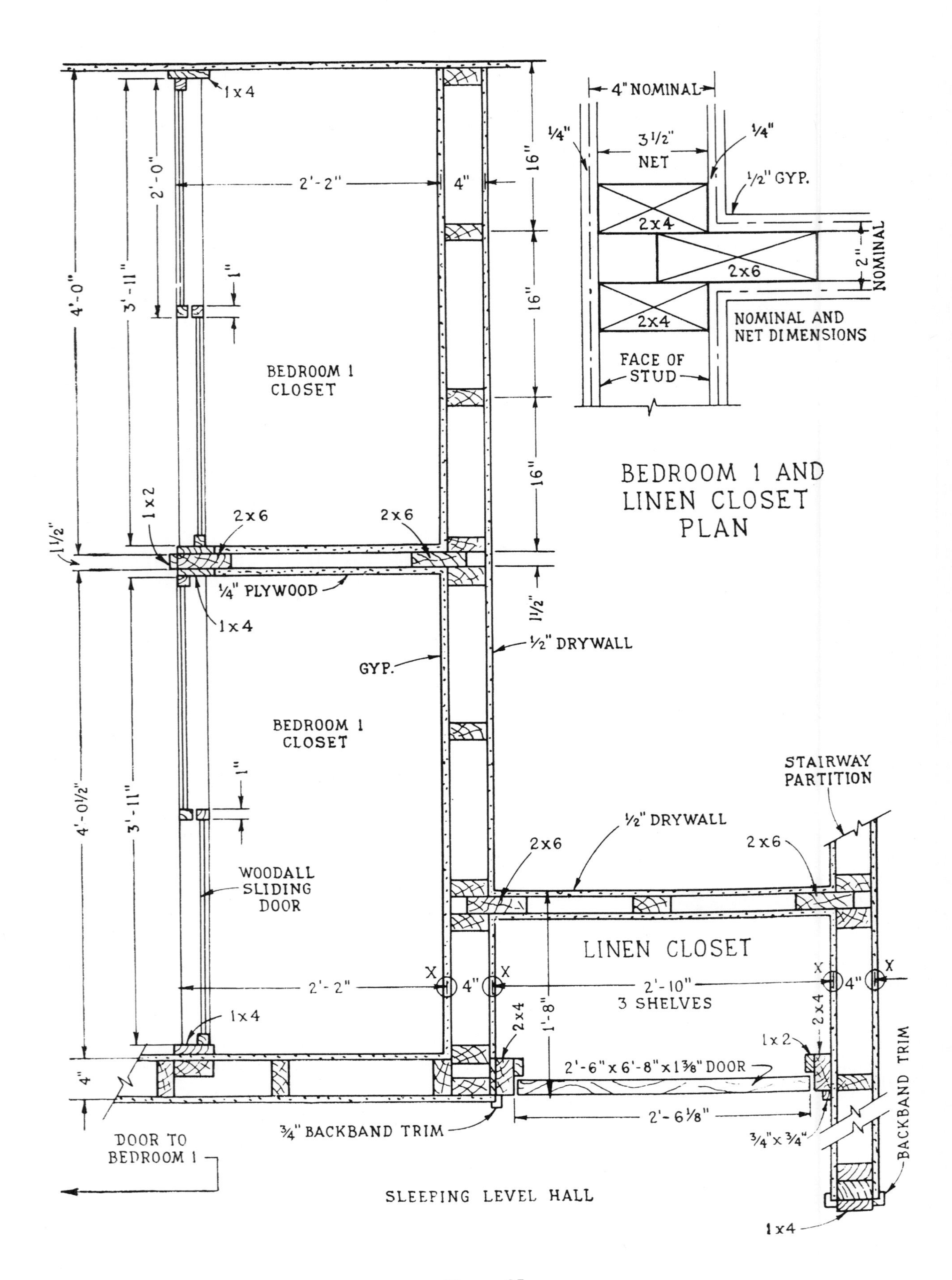
1x4
2'-0"
2'-2"
4"
16"
4'-0"
3'-11"
1"
BEDROOM 1
CLOSET
1x2
2x6
2x6
1 1/2"
1/4" PLYWOOD
1x4
1/2" DRYWALL
GYP.
BEDROOM 1
CLOSET
4'-0 1/2"
3'-11"
WOODALL
SLIDING
DOOR
2'-2"
4"
X
1x4
4"
DOOR TO
BEDROOM 1
3/4" BACKBAND TRIM
SLEEPING LEVEL HALL
4" NOMINAL
1/4"
3 1/2"
NET
1/4"
1/2" GYP.
2x4
2x6
2" NOMINAL
2x4
NOMINAL AND
NET DIMENSIONS
FACE OF
STUD
BEDROOM 1 AND
LINEN CLOSET
PLAN
STAIRWAY
PARTITION
1/2" DRYWALL
2x6
2x6
LINEN CLOSET
2'-10"
3 SHELVES
1'-8"
2x4
2x4
1x2
2'-6" x 6'-8" x 1 3/8" DOOR
2'-6 1/8"
3/4" x 3/4"
BACKBAND TRIM
1x4

Figure 45

closets in bedroom 1 and a linen closet in the hall are indicated. As previously explained, the framing for these closets must be carefully planned. This is another good example of the necessity for detail views.

In Figure 45, the closets are shown in the same position as in Figure 26*A*.

We can visualize Figure 45 by imagining that the closets have been cut horizontally, as for plan views, so that we can look directly down at the cut surfaces from a position just above them. In order to read the detail view, the following observations are listed.

1. The rear partition of the linen closet is to be framed with a 2 by 4 and 2 by 6s having their long sides parallel to the partition.

2. The partition between the bedroom-1 closets is also to be framed with the long sides of the 2 by 6s parallel to the partition.

3. Wherever the thin partitions meet full-sized partitions, 2 by 6s are to be used to form three member corner posts with two 2 by 4s.

4. The back wall of the bedroom closets is to be framed with 2 by 4s in normal positions and spaced 16 inches O.C.

5. Woodall sliding doors are specified. The framing around them is to be made with 1 by 4s.

6. The framing around the linen closet door is to be built using 2 by 4s and 1 by 2s. A ¾- by ¾-inch square molding and a ¾-inch backband are to be used as trim.

7. The end of the partition between the linen closet and the staircase is to be framed using two 2 by 4s, a piece of 1 by 4, and two pieces of backband trim.

8. Between the two openings for the bedroom closets, a 2 by 6, two pieces of 1 by 4, and a piece of 1 by 2 are to constitute the framing.

9. All closets are to be lined with gypsumboard.

At this point, we should consider the fact that 2 by 4s and 2 by 6s, for example, are not actually those dimensions. The true dimensions are closer to 1½ by 3½ inches and 1½ by 5½ inches. With this fact in mind, architects generally create dimensions which compensate for the difference between *nominal* and *actual* (net) dimensions. Note the small sketch which is labeled, "Nominal and Net Dimensions." We can see here the difference between nominal and net dimensions. The nominal width dimension for the 2 by 4 width is 4 inches. The net width is 3½ inches. The difference between them is ½ inch.

In order to compensate for the difference between nominal and net dimensions, architects extend the dimensions as shown at the *X* points in the planlike and sectionlike detail view. In the "Nominal and Net Dimensions" sketch, long-and-short dashed lines are drawn almost to the centers of the gypsumboard. We must imagine that similar lines are drawn in the gypsumboard at the *X* points in the detail view. In other words, designers base their dimensions on nominal widths and show the ends of the dimensions accordingly. This practice avoids the use of fractions of inches in the dimensions shown in construction plans.

As a further example relative to the actual thickness of 2 by 4s and 2 by 6s, note the piece of 1 by 2 trim between the two bedroom closets. The trim is actually 2 inches wide. Thus, it is shown as being wider than the net thickness of the 2 by 6 to which it is to be nailed.

Figure 46 This illustration constitutes another example of an assembly-class planlike and sectionlike detail view.

In the living-level part of Figure 26*A*, coat and broom closets are indicated near the kitchen symbol. The *A* part of Figure 46 shows these closets in the same position as shown in Figure 26*A*. This detail view must be visualized as though the closets have been cut through horizontally so we can look directly down at the cut surfaces. The view is read as explained in connection with Figure 45. Note that the 2′ 6″ coat-closet width checks with a like dimension in Figure 26*A*. Note, too, that more necessary dimensions are shown in the detail view.

The *B* part of Figure 46 shows an enlarged

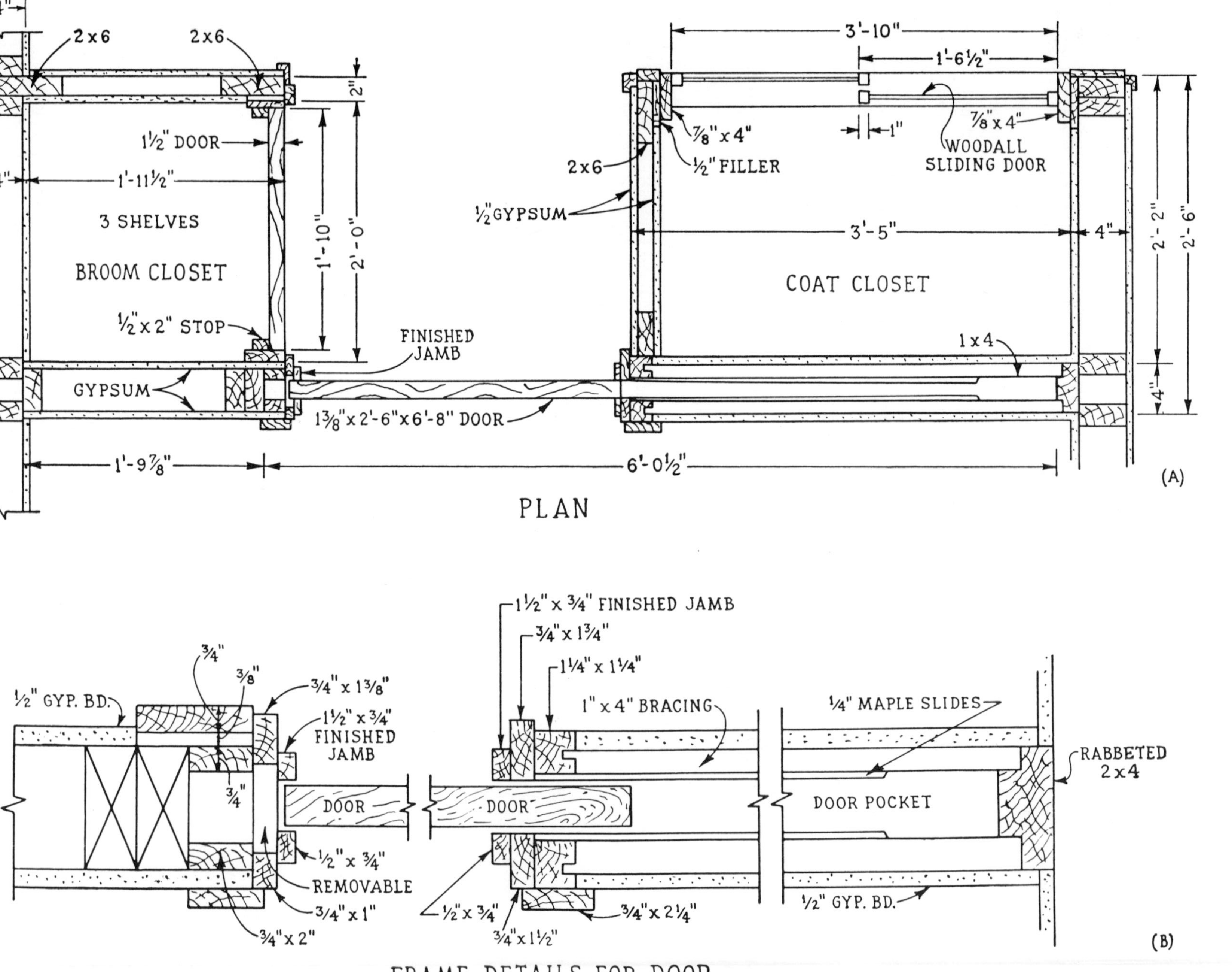

4"
2 x 6
2 x 6
2"
1½" DOOR
4"
1'-11½"
3 SHELVES
BROOM CLOSET
1'-10"
2'-0"
½" x 2" STOP
FINISHED JAMB
GYPSUM
1⅜" x 2'-6" x 6'-8" DOOR
1'-9⅞"
6'-0½"
3'-10"
1'-6½"
⅞" x 4"
1"
⅞" x 4"
WOODALL SLIDING DOOR
2 x 6
½" FILLER
½" GYPSUM
3'-5"
4"
2'-2"
2'-6"
COAT CLOSET
1 x 4
4"
(A)
PLAN
1½" x ¾" FINISHED JAMB
¾" x 1¾"
1¼" x 1¼"
1" x 4" BRACING
¼" MAPLE SLIDES
½" GYP. BD.
¾"
⅜"
¾" x 1⅜"
1½" x ¾" FINISHED JAMB
¾"
DOOR
DOOR
DOOR POCKET
RABBETED 2 x 4
½" x ¾"
REMOVABLE
¾" x 1"
¾" x 2"
½" x ¾"
¾" x 1½"
¾" x 2¼"
½" GYP. BD.
(B)
FRAME DETAILS FOR DOOR

Figure 46

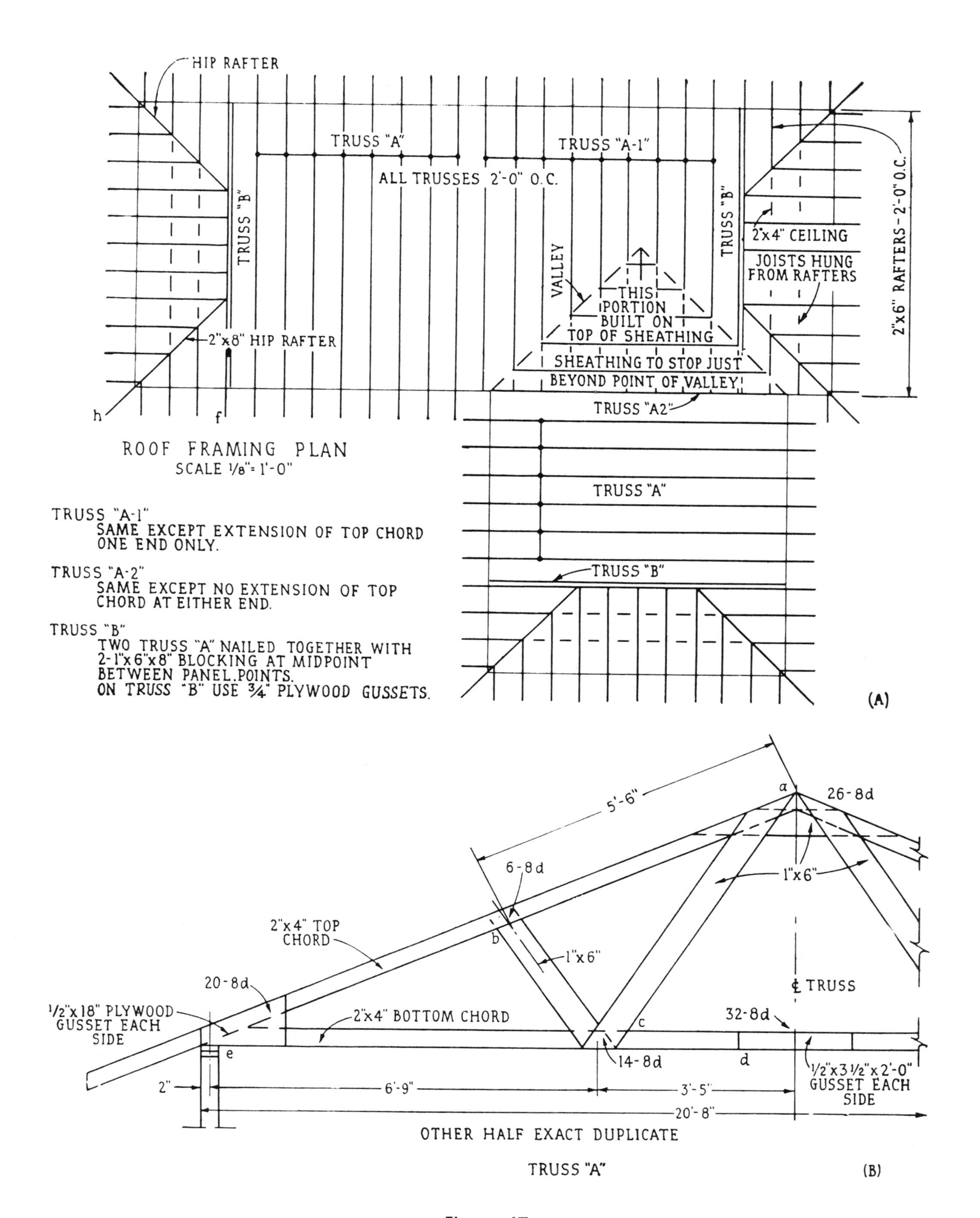
HIP RAFTER
TRUSS "A"
TRUSS "A-1"
ALL TRUSSES 2'-0" O.C.
TRUSS "B"
TRUSS "B"
VALLEY
THIS PORTION BUILT ON TOP OF SHEATHING
SHEATHING TO STOP JUST BEYOND POINT OF VALLEY
2"x4" CEILING
JOISTS HUNG FROM RAFTERS
2"x6" RAFTERS - 2'-0" O.C.
2"x8" HIP RAFTER
h
f
TRUSS "A2"
ROOF FRAMING PLAN
SCALE 1/8" = 1'-0"
TRUSS "A"
TRUSS "B"
TRUSS "A-1"
SAME EXCEPT EXTENSION OF TOP CHORD ONE END ONLY.
TRUSS "A-2"
SAME EXCEPT NO EXTENSION OF TOP CHORD AT EITHER END.
TRUSS "B"
TWO TRUSS "A" NAILED TOGETHER WITH 2-1"x6"x8" BLOCKING AT MIDPOINT BETWEEN PANEL POINTS.
ON TRUSS "B" USE 3/4" PLYWOOD GUSSETS.
(A)
5'-6"
a
26-8d
1"x6"
6-8d
b
2"x4" TOP CHORD
1"x6"
℄ TRUSS
20-8d
1/2"x18" PLYWOOD GUSSET EACH SIDE
2"x4" BOTTOM CHORD
c
32-8d
e
14-8d
d
1/2"x3 1/2"x2'-0" GUSSET EACH SIDE
2"
6'-9"
3'-5"
20'-8"
OTHER HALF EXACT DUPLICATE
TRUSS "A"
(B)

Figure 47

detail view of the sliding door to be built between the two closets. The pocket for the door and the jamb sections, for both sides of the door opening, are shown so that the size and position of all components are indicated.

Figure 47 This figure constitutes an example of a planlike and elevationlike detail view.

In Figure 12, the elevation views include a note which refers to details for roof trusses. This means that the roof of the one-story house is to be constructed by means of trusses rather than by the use of rafters.

The *A* part of Figure 47 shows a planlike detail view which indicates the kinds of trusses and spacings required for them and for the few regular rafters necessary at the three ends of the roof. Note that the shape of the drawing is the same as in the projected ceiling plan shown in the *A* part of Figure 41.

The *B* part of Figure 47 shows an elevationlike detail view of truss construction which we will study before the roof plan.

The detail view shows only half of the truss. However, the other half is exactly the same. In order to read the detail, the following observations are listed.

1. The pieces which are to be the top and bottom chords must be 2 by 4s.

2. The pieces between *b* and *c* and between *a* and *c* are to be 1 by 6s.

3. At the point where the 1 by 6 piece is to be attached to the top chord at *b*, six 8d nails are required. In other words, that joint is to be fastened with nails which are 2½ inches long.

4. Where the 1 by 6 piece is attached to the bottom chord at *c* together with the 1 by 6 piece from *a*, fourteen 8d nails are required.

5. At the ridge point *a*, a piece of 1 by 6 is to be used to hold the two pieces of top chord and the two pieces of 1 by 6 together. Twenty-six nails are to be used for that joint.

6. Where the top and bottom chords are to meet at *e*, a plywood gusset (connecting piece) having ½-inch (thick) and 18-inch dimensions is to be used. Twenty 8d nails are required.

7. Where the two pieces of bottom chord are to meet at *d*, a gusset having dimensions of ½ by 3½ inches by 2′ 0″ is required, to be fastened with 32 8d nails.

8. The center line of piece *bc* is to be 5′ 6″ from the center line of the truss.

9. The length of the truss from the outer edge of the plate of one wall to the outer edge of the plate for the opposite wall is to be 20′ 8″.

There are some changes necessary in order for trusses to be used at certain places in the roof. The notes relative to trusses *A1, A2,* and *B,* explain the changes.

In the *A* part of the illustration, the locations and spacings of the trusses are indicated.

The elevation views, in Figure 12, indicate that each of the three ends of the roof must have a part which slants up from the cornice. At the left-hand end of the detail view, the *hf* part of the roof constitutes the hip. To form that part of the roof, 2 by 8 hip rafters are required.

The bottom chords of the trusses are to serve as ceiling joists. Under the hip areas, the ceiling joists indicated by the dashed lines are to be hung from the rafters which, in turn, must be supported by the *B* trusses and the exterior walls.

Other information can be read from the detail view in much the same manner. Note that no dimensions are shown relative to the truss overhang. This is because the necessary cornice information is shown on the section view in Figure 35.

QUESTIONS AND ANSWERS

The questions in this chapter refer to Figures 46, 48, 49, and 50. However, it may be necessary to refer to Figures 26*A*, 26*B*, 29, and 30. Figure 46 shows a detail view of the coat and broom closets indicated in Figure 26*A*. Figure 48 shows a wall section in connection with the house represented in Figures 29 and 30. Figures 49 and 50 are detail views of the kitchen indicated in Figure 29.

As a means of making sure that you have learned to visualize detail views (and as a review of section views) and that you understand the relationship between them and other views, answer each of the following questions, either orally or in written form, and then check your reading with the *descriptive* answers shown.

Questions 1 through 15 are about Figure 46.

Question 1 How thick must the filler in the coat-closet framing be?
Answer 1 The filler must be the same thickness as the gypsumboard or ½ inch.

Question 2 What size finish jambs must be used to trim the sliding door opening?
Answer 2 There are two sizes necessary. One size is indicated as 1½ by ¾ inches and the other as ½ by ¾ inches.

Question 3 What size sliding door is required for the kitchen?
Answer 3 On the planlike and sectionlike view, the door is specified as 2′ 6″ by 6′ 8″.

Question 4 How many rabbeted 2 by 4s are required?
Answer 4 Where the door pocket meets the partition, one rabbeted 2 by 4 is required.

Question 5 How many sliding doors are required for the coat closet?
Answer 5 The planlike and sectionlike view indicates that two such doors are required.

Question 6 How wide is each of the coat-closet doors to be?
Answer 6 Each door is to be 1′ 6½″ wide.

Question 7 What type of door is required for the broom closet?
Answer 7 The detail view does not show a sliding-door symbol so we conclude that an ordinary hinge-type door is required.

Question 8 What size stops are required for the broom-closet door?
Answer 8 The stops are to be made of ½- by 2-inch material.

Question 9 Where are slides required?
Answer 9 Slides are required on both sides of the door pocket.

Question 10 How many 2 by 4s and 2 by 6s are required for the broom-closet framing?
Answer 10 Seven 2 by 4s and two 2 by 6s are required.

Question 11 Where is bracing to be used?
Answer 11 Bracing is to be used on both sides of the large sliding-door pocket.

Question 12 What size material is required for the finish framing around the coat-closet sliding doors?
Answer 12 The framing is to be of ⅞- by 4-inch material.

Question 13 Where is maple wood required?
Answer 13 For the slides in the door pocket.

Question 14 What is the interior width of the broom closet to be?
Answer 14 In order to determine that dimension, subtract 1½″ (door thickness) + ¼″ (half of gypsumboard thickness) from the 1′ 11½″ dimension. The exact width is to be 1′ 9¾″.

Question 15 What is the width of the broom-closet door to be?
Answer 15 The width is to be 1′ 10″.

Questions 16 through 29 are about Figure 48.

Question 16 What size main-floor joists are indicated?

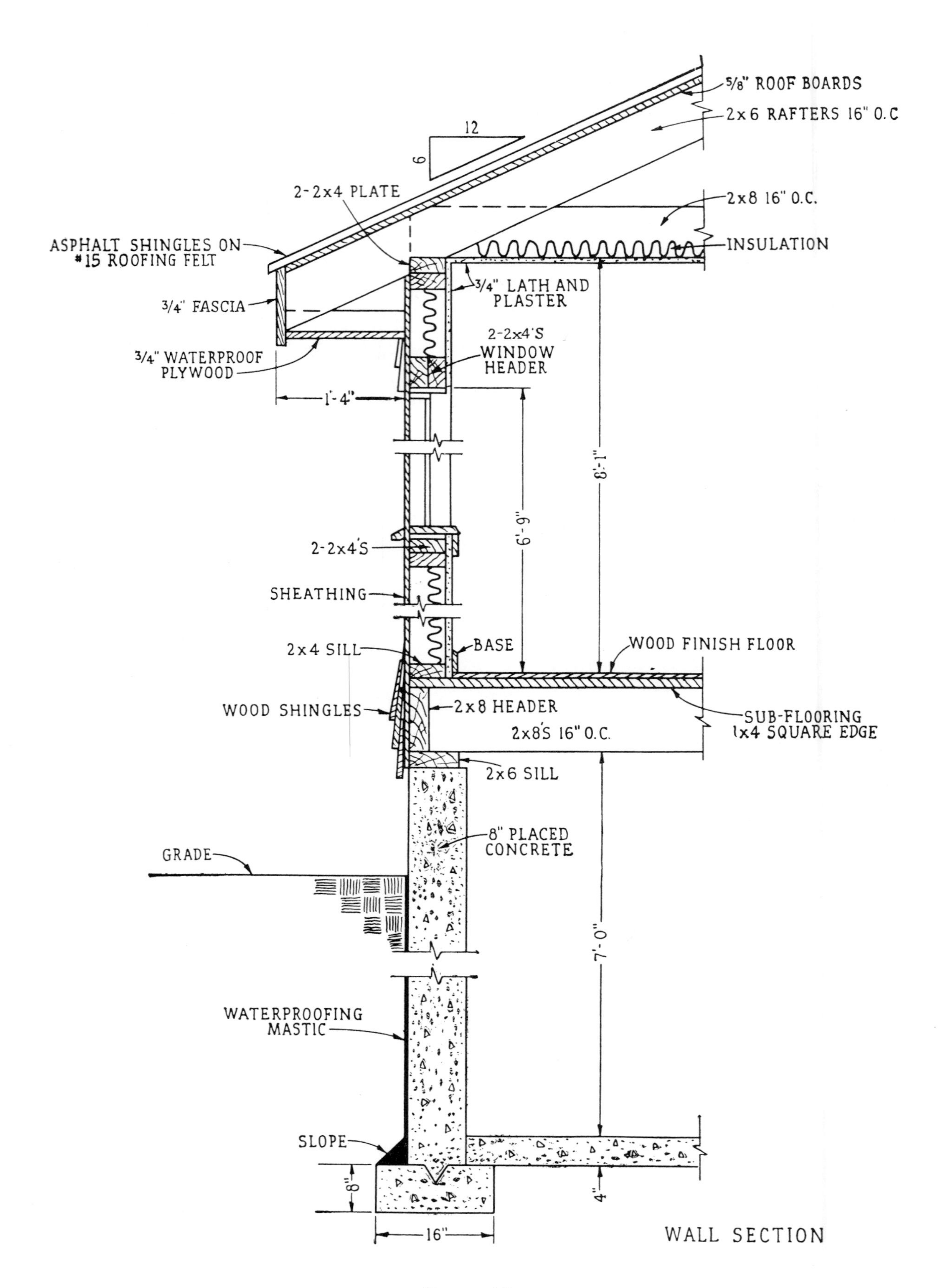
5/8" ROOF BOARDS
2x6 RAFTERS 16" O.C
12
6
2-2x4 PLATE
2x8 16" O.C.
ASPHALT SHINGLES ON
#15 ROOFING FELT
INSULATION
3/4" LATH AND
PLASTER
3/4" FASCIA
2-2x4'S
WINDOW
HEADER
3/4" WATERPROOF
PLYWOOD
1'-4"
8'-1"
6'-9"
2-2x4'S
SHEATHING
2x4 SILL
BASE
WOOD FINISH FLOOR
WOOD SHINGLES
2x8 HEADER
2x8'S 16" O.C.
SUB-FLOORING
1x4 SQUARE EDGE
2x6 SILL
8" PLACED
CONCRETE
GRADE
7'-0"
WATERPROOFING
MASTIC
SLOPE
8"
4"
16"
WALL SECTION

Figure 48

Answer 16 They are to be 2 by 8s.

Question 17 How high is the top of the foundation to be above the footing?
Answer 17 The height is calculated by adding the basement floor thickness less the thickness of the sill to the 7′ 0″ dimension. The floor thickness is 4″. The sill thickness is 1½″. Adding 2½″ to 7′ 0″ gives 7′ 2½″.

Question 18 What kind of subflooring is required?
Answer 18 The subflooring must be 1 by 4 square-edged boards.

Question 19 What size window header is required?
Answer 19 The header is to be made using two 2 by 4s.

Question 20 What is the exact distance from the surface of the floor to the window heads?
Answer 20 The exact distance is 6′ 9″.

Question 21 What is the exact distance from the surface of the floor to the top of the plate?
Answer 21 The exact distance is 8′ 1″.

Question 22 Where is insulation to be used?
Answer 22 Between studs and ceiling joists.

Question 23 What is the soffit to be made of?
Answer 23 It is to be made of waterproof plywood.

Question 24 Where is felt required?
Answer 24 Felt is required between the shingles and the roof boards.

Question 25 Where is mastic to be used?
Answer 25 It is to be used as waterproofing on the exterior side of the foundation below grade.

Question 26 Where should grounds be used?
Answer 26 They should be used between the baseboard and the studs.

Question 27 What thickness must the required plywood be?
Answer 27 At the soffit, it is shown as ¾ inch.

Question 28 What kind of shingles are required?
Answer 28 For the roof, asphalt shingles are required. For the exterior walls, wood shingles are required.

Question 29 What type of interior finish is required?
Answer 29 Lath and plaster.

Questions 30 through 35 are about Figure 49.

Question 30 What equipment is required for the kitchen?
Answer 30 The detail view shows that a range, an exhaust fan, a sink, a refrigerator, cabinets, and counters are required.

Question 31 Are wall cabinets required? If so, where?
Answer 31 Yes, wall cabinets are required along both the walls where the range and refrigerator are located.

Question 32 What widths are the range and refrigerator?
Answer 32 The range is 3′ 4″ wide and the refrigerator is 3′ 0″.

Question 33 How deep are the wall cabinets?
Answer 33 All cabinets are 12 inches deep overall.

Question 34 How wide are the counters?
Answer 34 They are 24 inches wide.

Question 35 What spacing is required between equipment?
Answer 35 There is a 2-inch space between the range and the sink and a 1-inch space between the counter and the refrigerator.

Questions from 36 on are about Figure 50.

Question 36 Where would we have to stand in the kitchen to see the elevationlike view shown at *A-A*?
Answer 36 We would have to stand in the center of the kitchen and look toward the range and sink.

Question 37 Is any furring to be done? If so, where?

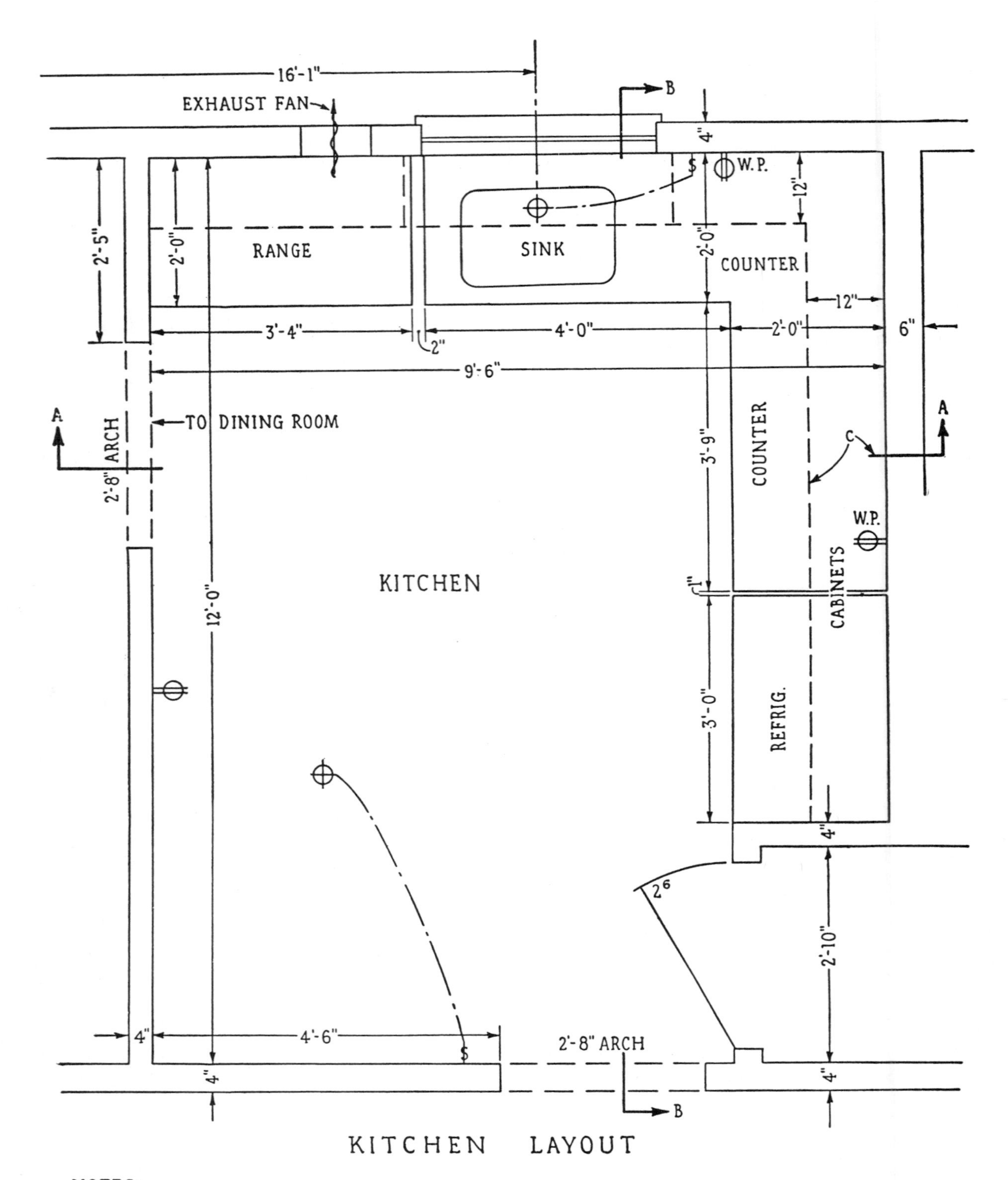

NOTES:
ALL DRAWINGS 1/2" = 1'-0" SCALE
ALL DIMENSIONS ARE FROM STUD, RAFTER OR JOIST LINE UNLESS OTHERWISE NOTED.

FOYER

Figure 49

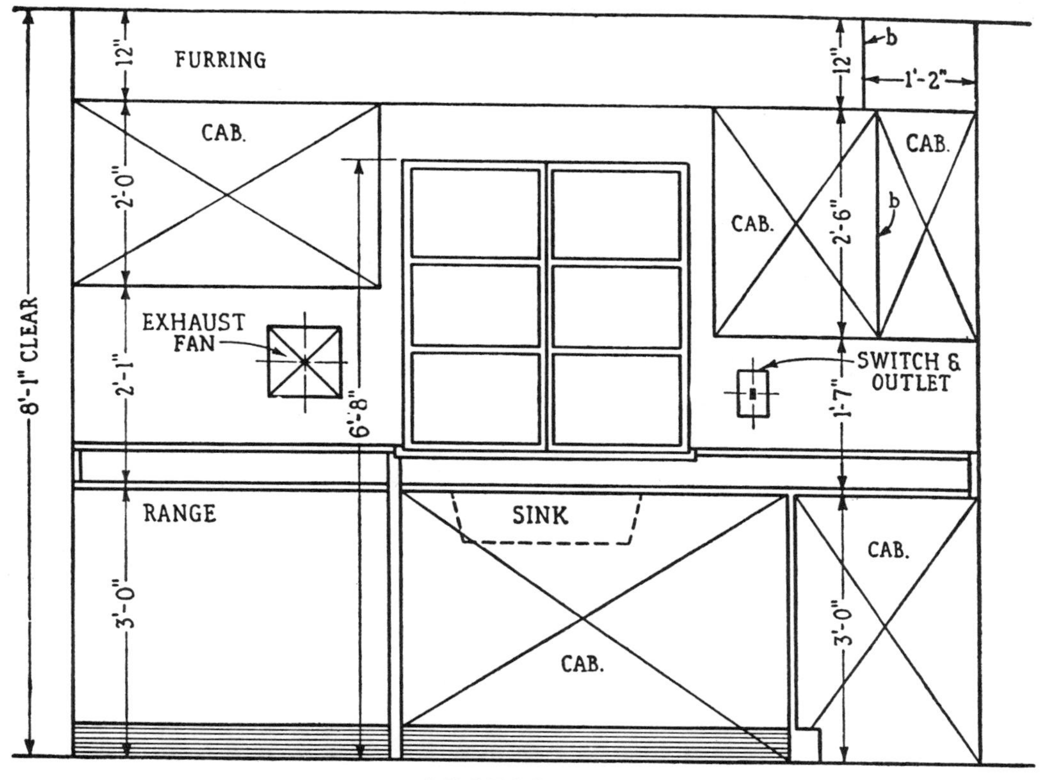

SECTION A-A

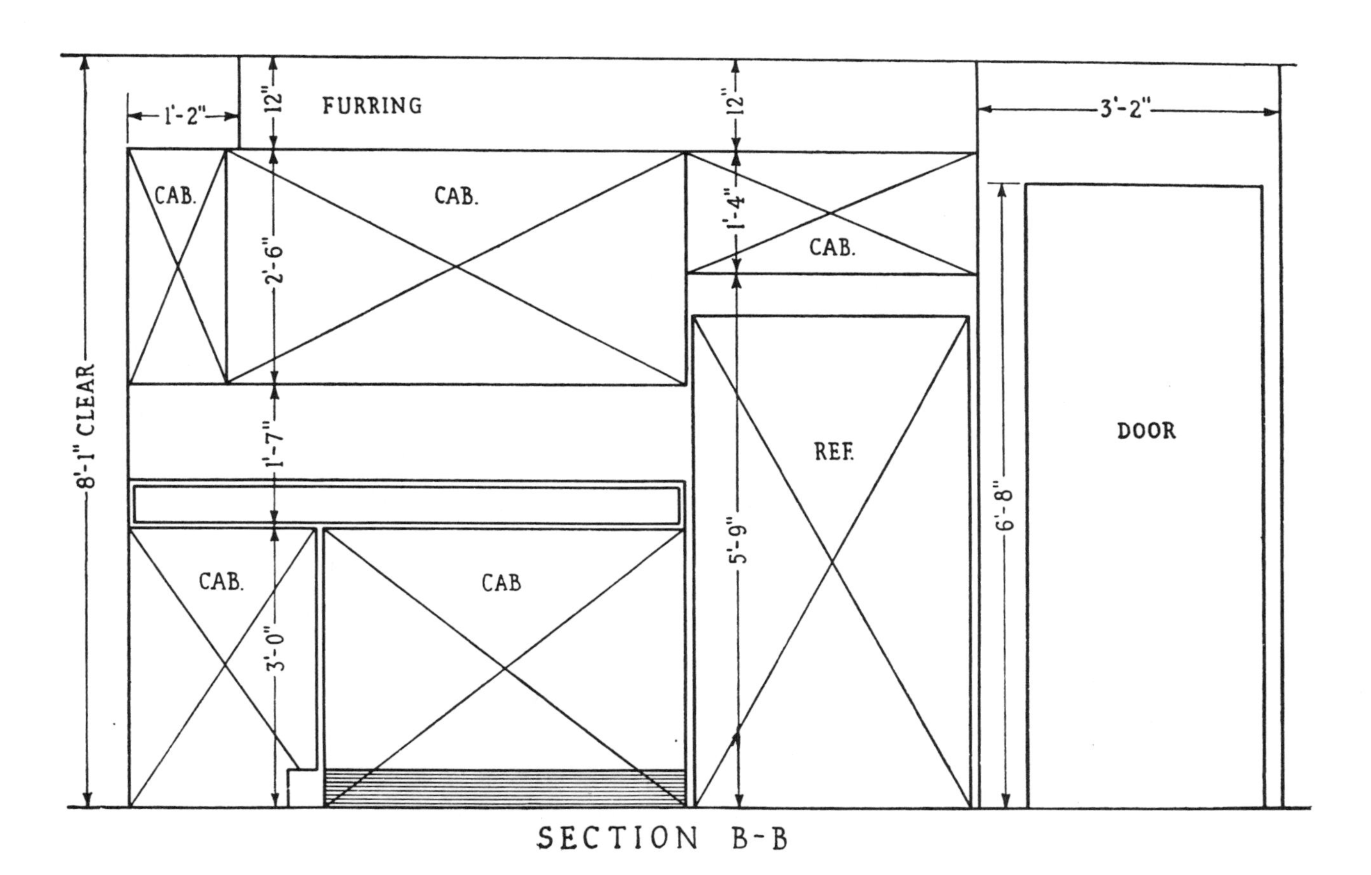

SECTION B-B

Figure 50

Answer 37 Yes. Furring is to be built along both of the equipment walls over the cabinets, sink, and refrigerator.

Question 38 How far above the counter tops are the bottoms of the wall cabinets?
Answer 38 The distance is 1′ 7″.

Question 39 How high are the counter tops above the floor?
Answer 39 Exactly 3′ 0″.

Question 40 What is the furring depth and width?
Answer 40 Both the *A-A* and the *B-B* views show that the depth is 12 inches and that the width is 1′ 2″.

Question 41 Is the furring width more than the overall depth of the wall cabinets?
Answer 41 Yes, 2 inches more.

Question 42 Which of the wall cabinets is to have the least height?
Answer 42 The cabinet over the refrigerator.

Question 43 Are the lines shown at *b* in the *A-A* detail actually section views?
Answer 43 Yes, because the cutting line *A-A* shown in Figure 49 cuts the cabinet as indicated at *c* in Figure 49.

Question 44 Where does the door shown in the *B-B* view lead to?
Answer 44 To the stairs which go down to the den level.

Question 45 How far is the top of the window above the floor?
Answer 45 The distance is 6′ 8″.

Question 46 How many electrical switches are required in the kitchen? Where?
Answer 46 Two are required. One is near the foyer arch and one is to the right of the sink.

Question 47 Are there wall cabinets above all the equipment?
Answer 47 Yes, except over the sink.

Question 48 Are the bottoms of all the wall cabinets 1′ 7″ above the counters and equipment?
Answer 48 No, over the range the bottom of the cabinet is 2′ 1″ above the range top.

Question 49 What is the clear distance between the floor and ceiling?
Answer 49 Exactly 8′ 1″.

Question 50 How far is the bottom of the smallest wall cabinet above the floor?
Answer 50 The distance is 5′ 9″.

CHAPTER SEVEN

Mechanical Details

Mechanical details are special drawings which include all the necessary information concerning electrical, heating, air-conditioning, and plumbing systems.

For *large* structures, such as schools, hotels, and office buildings, mechanical details are absolutely necessary, because of the complexity of the required systems, and are prepared in addition to the regular working drawings. Engineers and architects are mutually dependent, working out every detail carefully: architects, their plans and designs; engineers, their mechanical equipment.

For *small* structures, such as ordinary houses, mechanical details are seldom necessary because competent electricians and master plumbers are able to install the necessary equipment and connections, using the symbols shown on the regular drawings as a guide.

When preparing the working drawings for ordinary houses, most designers include the symbols concerning electrical and plumbing requirements. Such information can be included without adding too many confusing lines to the drawings. On the other hand, heating and air-conditioning symbols are seldom included because they would tend to cause confusion and because the installers generally prefer to plan their own work.

With the foregoing explanation in mind, we can understand that the drawings representing ordinary small structures seldom include mechanical details. Thus, the mechanics who work, or hope to work, on such structures will have little, if any, need to read other than the usual drawings which have been explained in the previous chapters of this book. However, in order that we will know what mechanical details are and how to read them, so far as ordinary small structures are concerned, the purpose of this chapter is to illustrate and explain typical examples of this type of detail. We shall depart somewhat from the format of the previous chapters.

HOW TO VISUALIZE MECHANICAL DETAILS

For the most part, mechanical details are visualized using the same procedure as explained for assembly-class detail views, in that a combination of planlike, and elevationlike and planlike, drawings are employed to show the necessary information.

Electrical As we learned in our study of Figures 26*A*, 26*B*, 27*A*, 27*B*, 28, 29, and 30, electrical symbols are shown on plan views. The symbol for ceiling lights, electrical outlets, switches, etc., must be visualized in their proper places. For example, a wall switch is generally installed about 54 inches above the floor and electrical outlets are generally in or just above baseboards.

Heating and Air Conditioning As for electrical layouts, heating and air-conditioning pipes and ducts are drawn on plan views. Thus, we imagine that we are looking down at the symbols from a point directly above them.

Plumbing For plumbing details, both planlike views and elevationlike views are sometimes drawn. In the planlike views, the pipes must be visualized as being in horizontal positions just under the floor. In the elevationlike views, we think of the pipes as being in vertical positions within walls and partitions.

SYMBOLS

The symbols employed for mechanical details include many we are already familiar with and a few new ones.

Electrical Typical electrical symbols are shown in Figure 23 and in the plan views in Figures 26*A*, 26*B*, 27*A*, 28, 29, and 30. In detail views, the dashed lines indicate the locations of switches and the outlets the switches are to control. The ceiling and convenience outlets are to be placed in the positions indicated by the symbols.

Heating and Air Conditioning Some of the symbols are shown in Figure 23. Radiators, ducts, and pipe symbols are indicated.

When forced-air heating and air-conditioning systems are involved, the horizontal runs of ducts in basements, crawl spaces, or between joists are indicated by solid parallel lines and by dashed lines. Generally, the solid lines indicate ducts which are complete by themselves and the dash lines indicate ducts which are to be made by panning the joists. That is a process of attaching sheet metal to the bottoms of adjacent joists so that the sheet metal, the joists, and the floor above constitute the duct. Arrows are used to indicate the direction of air travel.

Plumbing Some of the common plumbing symbols are shown in Figure 23. The symbols indicating water closets, lavatories, and bathtubs are fairly standard. However, pipe symbols vary. For that reason, most plumbing details include a key to symbols.

CONSTRUCTION-PLAN STUDIES

Figures 51, 52, and 53 represent actual mechanical details, as engineers prepare them and as they are used by the trades.

Figure 51 This illustration shows a typical electrical layout for one floor of a small home. Note that the plan-view symbols are shown only in enough outline to establish the rooms, walls, partitions, doors, and windows.

In order to read the electrical layout, the following observations are listed.

1. The garage light is to be controlled by two switches: one is to be located on the wall near the rear-hall door and one in the garage. This information is indicated by the S3 switch symbol.

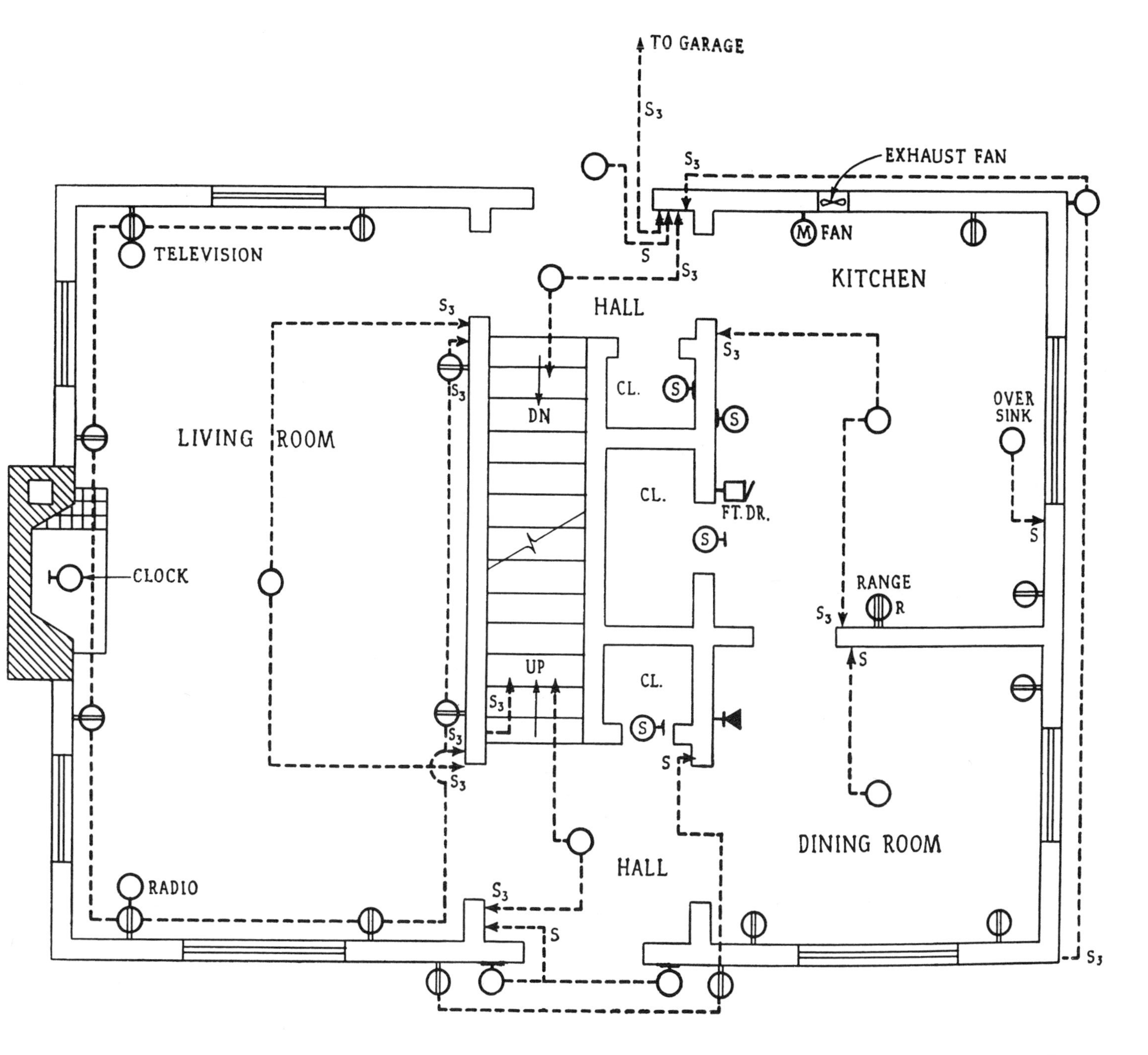
TO GARAGE
EXHAUST FAN
FAN
KITCHEN
TELEVISION
HALL
LIVING ROOM
CL.
DN
OVER SINK
FT. DR.
CLOCK
RANGE
UP
DINING ROOM
HALL
RADIO

Figure 51

2. Each of the closets is to have a wall-bracket-type light fixture and a pull switch. This is indicated by the T-like symbol attached to the circular symbol and by the letter S. In the rear-hall closet, the outlet is to be on the right-hand wall. In the other two closets, the outlets are to be over the doors.

3. The rear hall is to have a ceiling light so placed that both the hall and basement stairs will be lighted. This light is to be controlled by two switches: one near the rear exterior door and one in the basement.

4. The front hall is to have a ceiling light controlled by two switches: one in the wall near the front door and one on the second floor.

5. The living room is to have a ceiling outlet which can be controlled by two switches, one near each of the doors in that room.

6. The living room is to have eight wall outlets which can be controlled by two switches: one near each of the two doors.

7. Outside the front door, two wall-bracket-type lights are to be installed so that they can be controlled by a switch which must be located on the wall of the hall.

8. Two weatherproof wall outlets are also required on the exterior wall of the house: one on each side of the front door. The switch to control them must be located near the door of the hall closet.

9. A clock outlet is indicated for the fireplace mantel.

10. The telephone outlet is to be located in the dining room.

11. The front-door buzzer is to be located on a wall near the kitchen closet door.

12. An electric-range outlet must be provided in the position indicated in the kitchen.

13. A motor outlet for operating the kitchen exhaust fan is required.

14. An exterior wall-bracket-type light is required at the exterior corner of the house near the garage and it must be controlled by one weatherproof exterior switch, to be installed near the front corner of the house, and by one interior switch, to be located near the rear door.

15. An exterior wall-bracket-type light is to be placed above the rear door. The switch to control it must be on the wall near the door.

16. The kitchen and dining rooms are to have ceiling outlets which can be controlled by switches to be located on the wall near the door between these two rooms.

Figure 52 This illustration shows a typical duct layout for a heating and air-conditioning system in a small house. Note that the plan-view structural symbols are shown in outline form and only as a means of indicating the rooms, walls, partitions, windows, and doors.

In order to read the heating and air-conditioning layout, the following observations are listed.

1. The living room is to have two 12 by 6 inlets: one is to be located in the partition between closet 1 and the room, and the other in the left-hand partition of the room. This information is indicated by the two supply-duct symbols and by inward-pointing arrows.

2. One 30 by 6 air outlet is required in the living room and must be located in the front exterior wall under the windows. This information is indicated by the return-duct symbol and by the outward-pointing arrows.

3. Note that the mutual partition of bedrooms 2 and 3 is created by a *folding* partition. Bedroom 2 is to have one 12 by 6 air inlet which must be located in the short length of built-in partition between bedrooms 2 and 3. Bedroom 3 is to have one 12 by 6 air inlet which must be located in the partition between that room and the living room.

4. Bedrooms 2 and 3 must each have one 14 by 6 air outlet. Both of the outlets are to be located in the front exterior wall.

5. Bedroom 1 must have one 12 by 6 air inlet which is to be located in the partition between that room and closet 2. This bedroom must also have one 14 by 6 air outlet located in the rear exterior wall.

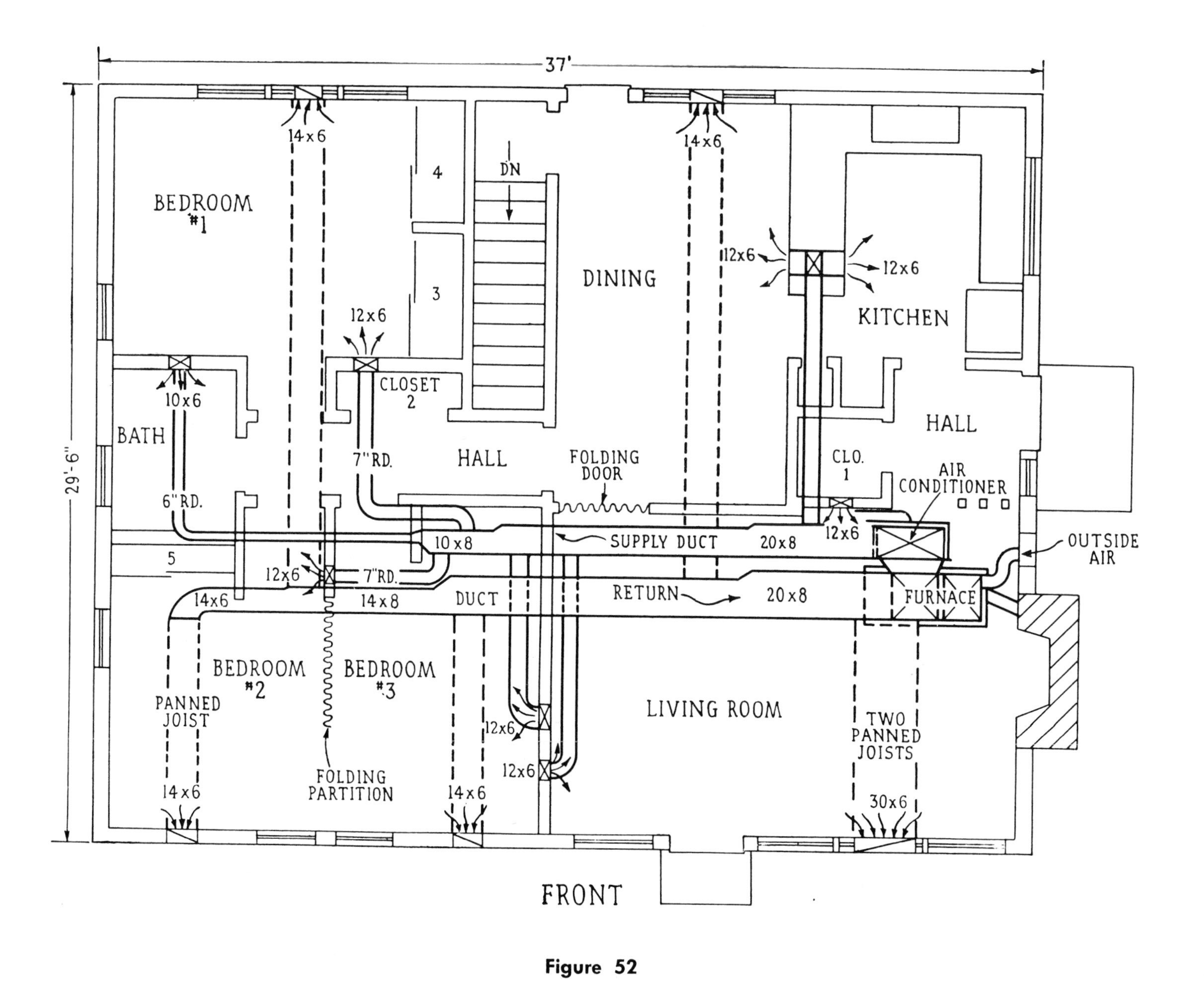
37'
29'-6"
14x6
14x6
DN
4
3
BEDROOM #1
DINING
KITCHEN
12x6
12x6
12x6
CLOSET 2
10x6
BATH
HALL
HALL
7"RD.
FOLDING DOOR
CLO. 1
AIR CONDITIONER
6"RD.
12x6
SUPPLY DUCT
20x8
10x8
OUTSIDE AIR
5
12x6
7"RD.
RETURN
20x8
FURNACE
14x6
14x8
DUCT
BEDROOM #2
BEDROOM #3
LIVING ROOM
PANNED JOIST
12x6
12x6
TWO PANNED JOISTS
FOLDING PARTITION
14x6
14x6
30x6
FRONT

Figure 52

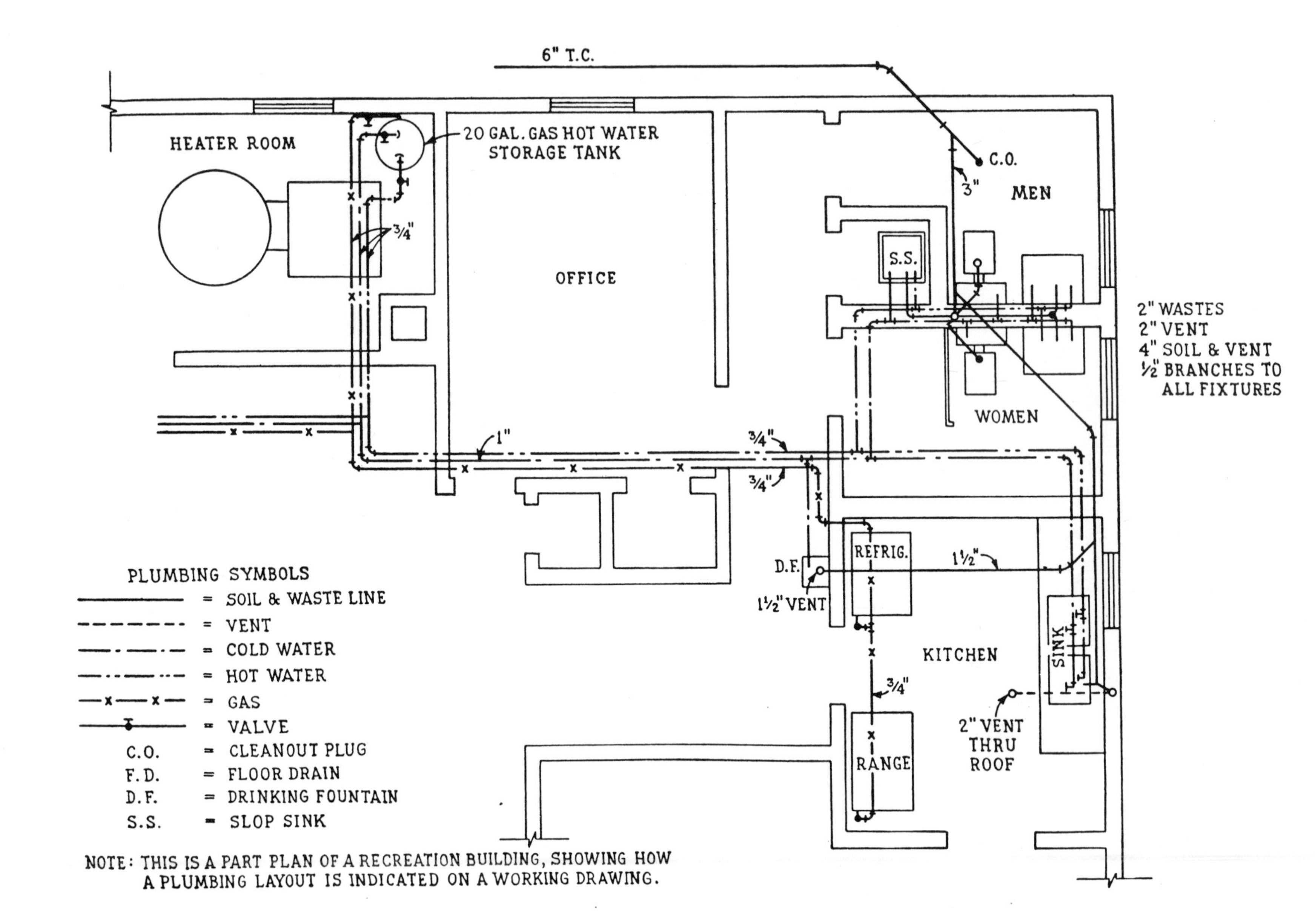
6" T.C.
HEATER ROOM
20 GAL. GAS HOT WATER STORAGE TANK
3/4"
OFFICE
C.O.
3"
MEN
S.S.
WOMEN
2" WASTES
2" VENT
4" SOIL & VENT
1/2" BRANCHES TO ALL FIXTURES
1"
3/4"
3/4"
D.F.
1 1/2" VENT
REFRIG.
1 1/2"
KITCHEN
SINK
3/4"
RANGE
2" VENT THRU ROOF
PLUMBING SYMBOLS
= SOIL & WASTE LINE
= VENT
= COLD WATER
= HOT WATER
= GAS
= VALVE
C.O. = CLEANOUT PLUG
F.D. = FLOOR DRAIN
D.F. = DRINKING FOUNTAIN
S.S. = SLOP SINK
NOTE: THIS IS A PART PLAN OF A RECREATION BUILDING, SHOWING HOW A PLUMBING LAYOUT IS INDICATED ON A WORKING DRAWING.

Figure 53

6. The dining room and kitchen must each have 12 by 6 air inlets. However, only the dining room is to have one 14 by 6 air outlet.

7. The main supply duct must be 20 by 8 where it connects to the air conditioner, and then gradually taper down to 10 by 8 where the last two branch ducts are connected to it.

8. The main return duct must be 20 by 8, where it connects to the furnace, and then gradually taper down to 14 by 6 where the first return-air duct is connected to it.

9. The branch supply ducts for bedrooms 1 and 2 are to be 7 inches in diameter.

10. The dashed lines indicate that the return air must travel between joists which have been panned.

11. The fact that most of the rooms have both supply and return ducts indicates that the air is to be recirculated. In other words, a considerable economy in operation costs can be realized by heating or cooling the air as it comes back from the various rooms instead of heating or cooling all fresh air.

12. There is to be a duct from the furnace to an exterior intake. This means that some fresh air can be mixed with recirculated air.

Figure 58 This illustration shows part of a typical plumbing layout for a small structure. Note that only enough plan-view symbols are shown to indicate the rooms, walls, and partitions. It should be pointed out that, to be complete, elevationlike details are necessary in order to show how the vertical pipes are to be placed and connected. This planlike view is ample for our purpose.

In order to read the plumbing layout, the following observations are listed.

1. The two lavatories, in the wash rooms, are to have both hot and cold water supplied by ½-inch branches.

2. The hot-water heater, the refrigerator, and the range must be supplied with gas through a ¾-inch pipe.

3. The drinking fountain is to be supplied with cold water and must have a 1½-inch drain.

4. The soil and vent stack must be 4 inches in diameter.

5. The house drain, from the soil stack to the sewer, must be made of 3-inch pipe.

6. The sewer must be made of 6-inch terracotta tile.

7. One cleanout is to be installed in the house sewer at the point where it is to be connected to the street sewer.

8. The water closets are to have ½-inch supply pipes.

9. The lavatories and the sink are to have ½-inch hot-water branches.

10. The main vent stack is to be located in the partition between the men's and women's rooms. A 2-inch vent, to the roof, is also necessary for venting the waste line from the kitchen sink.

11. The slop sink is to be connected to both hot and cold water.

12. Two-inch vent pipes are required for the slop sink, the lavatories, and the kitchen sink.

CHAPTER EIGHT

Survey and Plot Plans

A *survey* is a planlike view of a piece of land, showing its exact dimensions and levels, the positions and levels of existing trees, the lot boundaries with relation to existing streets or highways, the location and level of existing sewers and water mains, electric light and gas service, etc. Such plans must be made by licensed land surveyors who are authorized to do such work by cities or counties and whose plans are accepted as correct when issued and signed by them.

Most cities and counties require that a survey be made of any building lot or land before ownership can be established in any person's name. Surveys are also required before loans or mortgages can be secured for a proposed house.

A *plot* plan is also a planlike view which a designer prepares using a survey plan as a guide. On a plot plan, the exact position of a proposed structure is shown together with other information which builders must know before they can stake out the structure, erect batter boards, carry on excavation, build foundations, etc. A plot plan is also necessary as a guide for preparing finished grade after a structure has been constructed.

From the foregoing explanations, we can understand that both survey and plot plans are important to builders before and after construction is started. The purpose of this chapter is to illustrate and explain typical examples of such plans. As in Chapter 7, the following format differs from most of the previous chapters.

HOW TO VISUALIZE SURVEY AND PLOT PLANS

In a previous chapter we learned that all the views constituting a set of drawings for a proposed structure are referred to as *plans*. The same general terminology is applied to surveys and plots. However, both surveys and plots can be classed as planlike views because we must visualize them as though we are looking straight down at them from a position directly above them. Or we can imagine that the symbols used on such drawings can be drawn on the surface of the ground using the same sort of marking devices as used to lay out football gridirons, or to indicate lanes and parking spaces on streets and highways.

On plan views, the shapes of houses and parts of them are indicated by symbols in such a manner that we can see the necessary information. On survey and plot plans, symbols are used for the same purpose except that they show different information.

SYMBOLS

On both survey and plot plans, the symbols employed consist mostly of types of lines and a few of those symbols we learned in connection with plan views. It should be pointed out that neither survey nor plot is meant as a guide to structural items. Therefore, only such symbols as are necessary to indicate locations and shapes are shown.

Site Boundary Lines The exact boundaries of a piece of property, which is referred to as *lot* or *site,* are indicated by a line with a series of long dashes followed by two short dashes.

Contour Lines All contours are indicated by medium-weight solid lines which curve according to levels. This will be explained a little later in this chapter.

Public Utilities Sewers, water mains, etc., are indicated by a line composed of a series of short dashes.

Concrete Massed dots are used to indicate concrete surfaces such as driveways and walks. Solid lines are employed to outline such items. Other building materials are indicated by the same symbols we learned in connection with plan views.

Fills When excavated earth is to be used to fill or change the grade of a lot, dashed lines are employed to indicate the extent of the fill.

Structures A crosshatching, similar to the brick symbol shown on plan views, is used to indicate the shape and the position of a structure on such plans.

Trees The + (plus) sign is used to indicate the positions of all existing trees.

Notes On both survey and plot plans, considerable information can be supplied by means of printed notes.

CONTOURS

In Figures 54 and 55 there are several curved lines which have figures shown in connection with them. Such lines are known as *contours.* They are used to indicate the elevation of the different parts of a site. Here, the term *elevation* is used to indicate a height. In other words, we use the term to indicate how much higher various parts of a site are than some established height.

Most cities and towns have one or more points of known elevation or height above sea level. Such points are known as *bench marks.* In other cases, cities and towns establish what is called a *datum* which is given a certain elevation or height, such as 100 or 0. In either case, surveyors use the established elevation as a point from which to determine the elevations of a site. Here, for example, we shall assume that a bench mark of 100 feet above sea level is being used.

By use of an instrument called a *level,* surveyors can find and mark the elevations of all parts of a site. What they actually do is determine how far above or below the established bench mark each part of a lot is.

Let us suppose that Lot 114, shown in Figure 54, has a definite slope. In order to show the nature of that slope, a surveyor draws contour lines on his survey plan. Note the contour line

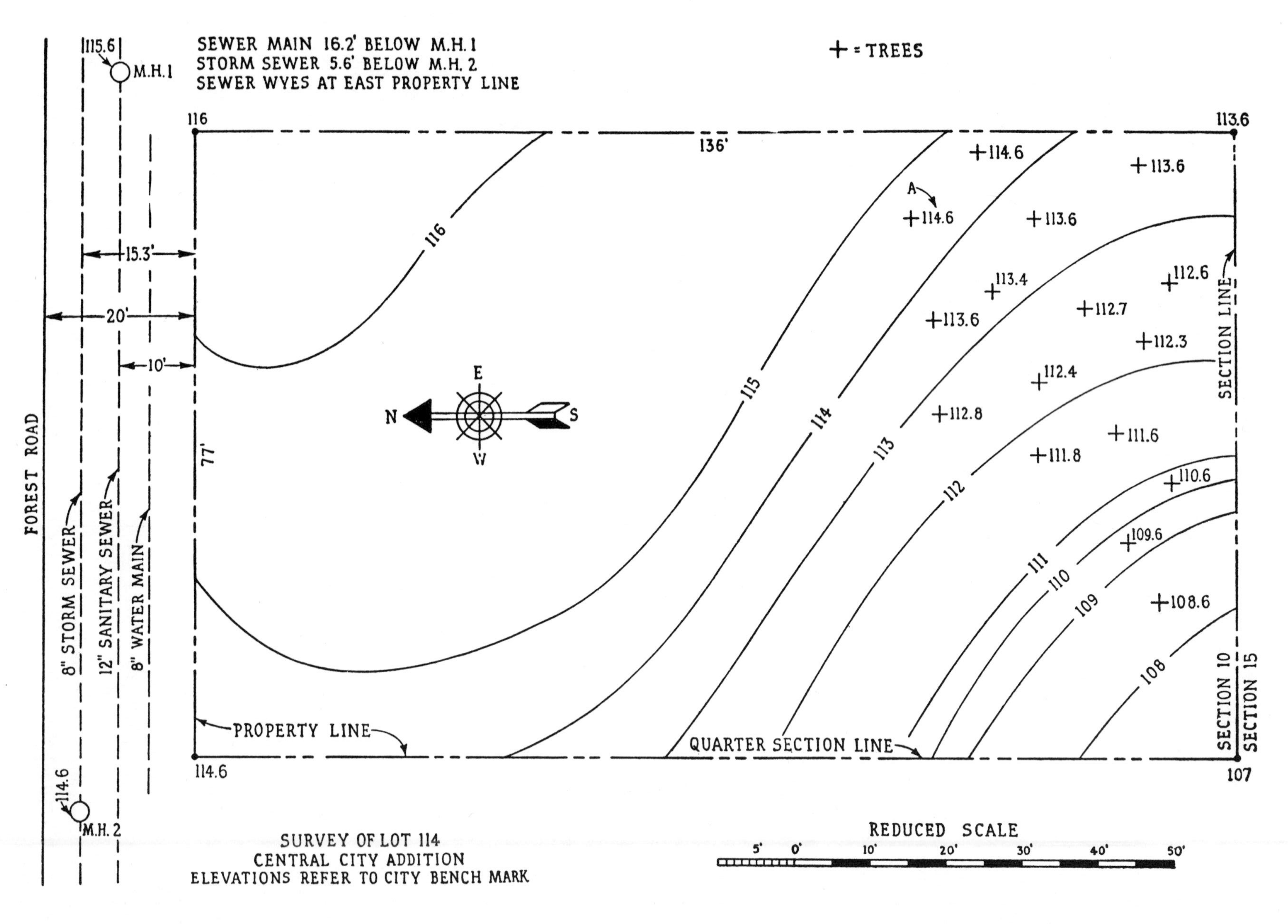
SEWER MAIN 16.2' BELOW M.H.1
STORM SEWER 5.6' BELOW M.H.2
SEWER WYES AT EAST PROPERTY LINE
+ = TREES
115.6
M.H.1
M.H.2
114.6
FOREST ROAD
8" STORM SEWER
12" SANITARY SEWER
8" WATER MAIN
15.3'
20'
10'
116
113.6
114.6
107
136'
77'
PROPERTY LINE
QUARTER SECTION LINE
SECTION LINE
SECTION 10
SECTION 15
N
S
E
W
A
116
115
114
113
112
111
110
109
108
+114.6
+114.6
+113.6
+113.6
+113.6
+113.4
+112.6
+112.7
+112.3
+112.4
+112.8
+111.6
+111.8
+110.6
+109.6
+108.6
SURVEY OF LOT 114
CENTRAL CITY ADDITION
ELEVATIONS REFER TO CITY BENCH MARK
REDUCED SCALE
5'
0'
10'
20'
30'
40'
50'

Figure 54

marked 116. All along that line, the surface of the site is at an elevation of 116 feet or 16 feet above the established 100-foot level at the bench mark. (If parts of a site are below a bench mark designated 0, the figures with the contours there are minus.) In like manner, all along the contour marked 115 the site is at an elevation of 115, or 15 feet above the established bench mark. From contour 116 to contour 115, there is a slope of 1 foot. The distance between contours indicates the part of the site which gradually slopes from 116 to 115 feet elevation. Where the contours are closer together, they indicate that the slope is less gradual. In other words, the closer contours are together, the steeper the slope. Between contours, the elevations are marked in tenths of a foot if some items, such as trees, are to have their bases indicated in terms of elevation.

DIMENSIONS

In making survey plans, surveyors use a special scale which is divided into feet and tenths of a foot. In other words, instead of showing distances of less than a foot in terms of inches, they use tenths of a foot. For example, 10.5 feet is used instead of 10′ 6″. The 0.5 is half of 1.0 just as 6 is half of 12. Or, as a further example, the surveyor's elevation of 117.3 feet is 117′ 3⅝″ when converted to the foot and inch scale. Builders seldom, if ever, have to work directly with a survey plan because designers prepare plot plans on which dimensions are shown in feet and inches.

SCALE

Surveyors generally use a scale in which 1/10 of a foot equals 1′ 0″. All parts of such plans are drawn accurately to scale.

Designers generally use a scale of 1/16″ = 1′ 0″ for plot plans. All parts of the plans are drawn accurately to scale.

SURVEY PLAN

Figure 54 shows a typical plan for a city lot. In order to read the survey plan, the following observations are listed.

1. The site is 136 feet long and 77 feet wide.
2. The rear end of the site is bounded by the line between sections 10 and 15.
3. One side of the site is bounded by a quarter-section line.
4. The site slopes from an elevation of 116 feet at its N.E. corner, down to 107 feet at its S.W. corner. In other words, the site slopes to the S.W.
5. The front half of the site is almost flat and thus provides an excellent location for a building.
6. The south and southeast ends of the site contain many trees. For example, the base of the tree marked *A* is at an elevation of 114.6 feet.
7. The storm sewer is located 15.3 feet beyond the front lot line.
8. Sewer connections (wyes) are in line with the east site line.
9. The sanitary sewer is located 16.2 feet below the top of manhole 1.
10. The street directly in front of the site has a slope of one foot between the east and west boundaries of the site.

PLOT PLAN

Figure 55 shows a typical plot plan which was drawn using the survey plan in Figure 54 as a guide.

In order to read the plot plan, the following observations are listed.

1. When the building is constructed, its finished floor level must be at an elevation of 117 feet.
2. The front of the structure is to be 20′ 0″ back from the front boundary line.

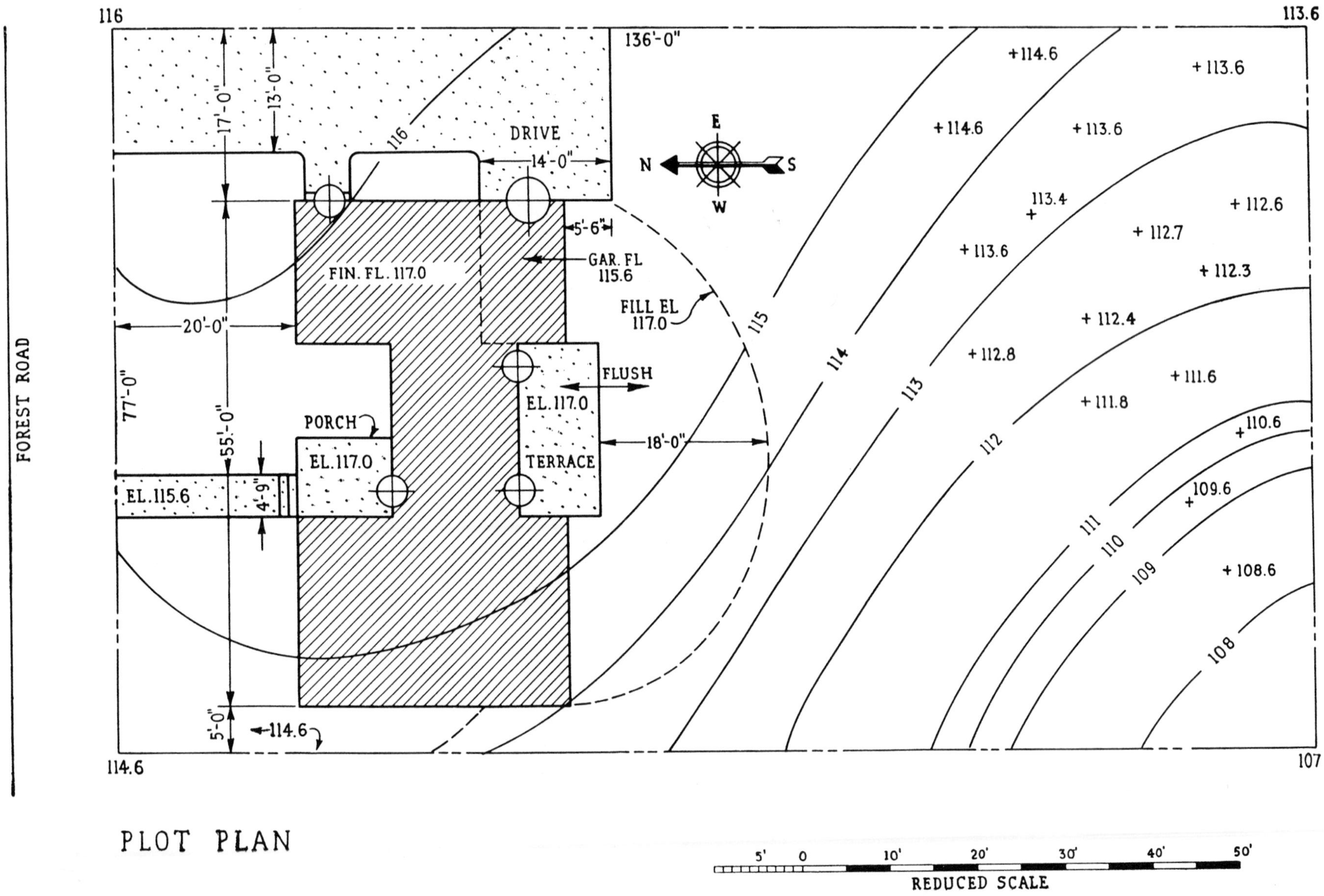

PLOT PLAN

Figure 55

3. The west end of the structure must be 5′ 0″ from the west boundary line.

4. The driveway is to be 13′ 0″ wide and is to extend 5′ 6″ beyond the S.E. corner of the structure.

5. The sidewalk, between the structure and the street, must be 4′ 9″ wide.

6. The elevation of the garage floor must be 115.6 feet.

7. The dashed line and the 18′ 0″ dimension indicate that a fill is required. Earth from the excavation is to be used to make the fill. The arrow and the word "flush" mean that the elevation of the fill is to be at the same elevation as the terrace floor. The dashed line shows the extent of the fill.

8. The driveway must have a slope of 116 minus 115.6, or 0.4 of a foot, between the street and the garage.

9. The finished floor, at the N.W. corner of the structure, is to be 117 minus 114.6, or 2.4 feet above the surface of the ground.

10. The finished floor, at the N.E. corner of the structure, must be 117 minus 116 or 1 foot above the surface of the ground.

11. The sidewalk is to be at elevation 115.6. Thus, it is to be 117 minus 115.6, or 1.4 feet lower than the porch floor. Steps are indicated.

CHAPTER NINE

Written Specifications

As previously explained, specifications are a written description of the work to be done in building a structure. In general, they cover all the features of the work not shown on the construction plans and which designers prefer to describe in writing. For example, the specifications describe the general conditions and stipulations, the quantity and quality of material, the methods of construction, the nature or standards of workmanship, and the manner of conducting the work. They are presented in typewritten form to accompany the drawings and must be read by all concerned parties.

Specifications include both general and specific explanations, the former covering the general conditions or stipulations that are to be used, with variations, for all types of construction. The specific explanations cover the nature of the workmanship, the quality or quantity of materials, the method of construction, and any special features pertaining to a particular structure.

To include in this book a complete set of specifications for a large and costly structure, which would include every possible detail of construction and equipment, would be impractical because such a complete specification would be a book in itself. It would also be unnecessary because if the fundamentals are shown, as in the specifications in this chapter, readers will grasp the general idea of what specifications are, something of how they are prepared, and how they are used.

The general specification form included with this chapter was not written to accompany a specific set of drawings for a particular structure. Instead, as its name implies, it is a specification which can be adapted to the needs of any set of drawings in connection with any ordinary small house.

The purpose of this chapter is to explain the fundamentals of specifications and to show how general specification forms are commonly used in connection with the plans for ordinary small houses.

PURPOSE OF SPECIFICATIONS

The foregoing definition of specifications answers the question as to what they are and what they contain. In the following, the principal purposes of them are set forth.

1. They are the means of avoiding disputes between builders, designers, and owners. They leave no room for misinterpretation or misunderstanding.
2. They facilitate the checking of materials and workmanship during the time a structure is being constructed.
3. They enable contractors to make bids based on total material and labor required.
4. They avoid many chances of expensive omissions when estimating total costs.
5. They avoid conflicting opinions and the added costs and delays which could result. For example, no two carpenters are likely to have exactly the same ideas as to what constitutes good workmanship or proper practice.

DIVISION OF SPECIFICATIONS

A specification for any ordinary structure should, in addition to a contract form and general and supplementary conditions, contain the following divisions:

1. General requirements
2. Site work
3. Concrete
4. Masonry
5. Metals
6. Wood and plastics
7. Thermal and moisture protection
8. Doors and windows
9. Finishes
10. Specialties
11. Equipment
12. Furnishings
13. Special construction
14. Conveying systems
15. Mechanical
16. Electrical

The general specification form shown at the end of this chapter is divided into such sections.

HOW TO USE SPECIFICATIONS

Both specifications and drawings may indicate information about the same items. For example, elevation views may show that asphalt shingles are to be used and the specifications may contain the same information. However, the two methods do not overlap to a great extent and no confusion results.

The symbols and notes appearing on an elevation or plan view, for example, may completely describe a certain aspect of construction. In that event, the specifications need not repeat the same information. When, however, symbols and notes do not contain complete information, then specifications must include whatever additional information may be required. For example, section and detail views may indicate that subflooring is required. That is not enough information. So, the specifications would have to show the size of boards to use, how the boards are to be mill-finished (such as *S4S*), and how nailed. In other words, if the drawing symbols and notes do not give complete information, we look in the specifications for whatever additional information we need.

As previously indicated, specifications are written in several divisions, each of which applies to a trade. For example, there is sure to be a division on wood work. In this division are given all the instructions which concern the carpentry work. Thus, a carpenter turns to the division on wood to find whatever additional information he needs. The same is true for the other trades. It is a good idea, however, for every mechanic to read the entire set of specifications in order to be familiar with the job as a whole and the general character of the requirements.

When we study drawings and specifications, we should pay careful attention to what is shown on the drawings, as well as what is set forth in the specifications. In other words, we cannot possibly obtain a full understanding of all requirements until we are thoroughly familiar with both plans and specifications.

OUTLINE SPECIFICATION FORM

The specification form, illustrated in the following pages of this chapter, is also known as an *outline* form of specification. That name originated because such forms are much more brief than the detailed forms sometimes used in connection with costly and out-of-ordinary houses or other structures.

Although in a brief form, the outline specification gives ample information to obtain bids, let contracts, safeguard both contractor and owner, and build the structure. This form can be and is used to advantage in connection with ordinary structures where most of the materials are stock items.

Divisions The various divisions of the specifications are prepared so as to meet the requirements of most building needs. In many instances, the instructions are given for more than one type of construction or equipment. The following examples illustrate the situation.

Siding The wood and plastics division of the following specifications contains instructions for both horizontal and vertical siding. If only horizontal siding is indicated on elevation views, the instructions for vertical siding can be ignored or crossed out.

Caulking The wood and plastics division also gives instructions for caulking at places where wood and masonry meet. If the elevation and plan views do not indicate masonry walls, the caulking instructions can be ignored or crossed out.

Special Needs Now and then an architect may prefer to give instructions about windows and doors in the specifications. In such cases, items not applicable in the doors and windows division can be struck out and new items required can be added. In like manner, other special instructions can be given in connection with any of the structural work or equipment.

Selection Many of the divisions in the general form contain one or more cases where the owner has to indicate a choice. For example, the owner may have to decide what type and price plumbing fixtures he wants. Or he may have to select the type of heating he wants. In such cases the specifications are marked accordingly.

Finishes The finishes division of the specifications requires an owner to decide on finishes for various rooms. A designer may help him make the selections and adjust or write the specs.

In general, we can expect that general specification forms will be modified to suit our needs as mechanics. When no modifications are shown, we use the specifications which apply to the type of construction equipment shown on the plans.

If we encounter more elaborate specifications there is no need to be concerned because they are also set up in divisions and the only difference is that they give more detailed instructions on nonstock items.

GENERAL SPECIFICATION FORM

GENERAL NOTE

This specification contains many items for selection of materials and methods, which must be completely filled in, in order to make this a true and complete specification. The reason for this is that the availability of materials, use of materials, and cost of materials vary in every locality.

Your selection will directly affect the cost of your building. All such items when selected shall become part of the contract and shall be mutually agreed upon by you and your Contractor.

CONTRACT

WE, THE UNDERSIGNED PARTIES TO THIS AGREEMENT, have read the following specifications and have accepted them as correct.

IT IS CLEARLY UNDERSTOOD and agreed by and between the parties hereto that the foregoing specifications are hereby made a part of a certain building contract of this date and covering the same building and that the said contract is accepted and based upon furnishing and installing in a First Class workmanlike manner all the items listed in the above specifications, and that the work will proceed with all haste consistent with good workmanship, weather, labor, and other conditions permitting.

ADDITIONAL ITEMS:

Total Cost as per Plan No. ______________ and Specification is

______________________________________ Dollars.

Dated ______________ 19____ CONTRACTOR ______________

Witnesses: ______________

______________ OWNER ______________

______________ ______________

FILE:
SET:

SYNOPSIS SPECIFICATIONS: For labor and materials required for the erection and completion of a residence for:

____

To be built upon Owner's property, located as described below:

____

____

A copy of the Plot Plan shall become part of these contract documents.

GENERAL REQUIREMENTS

1. NOTE: All work shall conform to minimum standards of the Regional Office of The Federal Housing Administration.

To apply to all Contractors and Subcontractors.

2. GENERAL: These specifications and the accompanying drawings are intended to describe and provide for a finished piece of work, and what is called for by either is as binding as if called for by both. The Contractor shall understand that the work described herein is to be complete in every detail, even though every item necessarily involved is not particularly mentioned.

3. CHANGES: Any changes in the plans during the progress of the work can be made without voiding the contract. However, the cost or allowance, as the case may be, of such changes, must be described and priced in writing and signed by the Owner and the Contractor.

4. LIABILITIES AND INSURANCE: Each Contractor shall be held responsible to maintain such insurance as will protect him from claims of workmen's compensation acts, or from any other claims for damages for personal injury, which may arise from the performance of his contract.

The Owner shall pay for fire insurance on the building during progress of the work. In case of fire, the Owner and the various Contractors will be paid according to the damages to their respective interests.

5. PAYMENTS: 85 percent of the value of labor and material satisfactorily incorporated in the work or stored at the site, will be made monthly when certified to by receipted bills submitted by the Contractor.

Final payment of 15 percent to be made within ten days after satisfactory completion and acceptance by the Owner. Payments will not be made until the Contractor delivers to the Owner a complete waiver of lien for all work executed.

Method of payments may vary with mortgagee; such variation shall be so stipulated in contract.

6. PAYMENTS WITHHELD: The Owner may withhold payments due, until the objection is remedied, for the following reasons:

(a). Defective work.

(b). Claims filed, or evidence indicating probable filing of claims.

7. MATERIALS, APPLIANCES, EMPLOYEES: Unless otherwise noted, the Contractor shall provide and pay for all materials, labor, water, tools, equipment, etc., necessary for the execution of the work and its completion.

Unless otherwise specified, all material shall be new and all workmanship and materials shall be of the best quality.

8. PERMITS AND REGULATIONS: The Contractor shall provide all required permits for work. The Contractor shall execute all work to conform to Local Building Code and F.H.A. specifications.

9. PROTECTION: The Contractor shall provide and maintain adequate protection of all his work from damage and shall protect the Owner's property from injury or loss during the period of the execution of the contract. He shall further protect adjoining property as required by ordinances and these documents.

10. SUPERINTENDENT: The Contractor shall keep on the work during its progress, a competent foreman, who shall be familiar with all drawings and specifications and who shall be responsible for the proper cooperation between the various Subcontractors.

11. SUBCONTRACTS: The Contractor may sublet any portion of the work involved to the extent of a minimum of two trades on his own regular payroll. He shall be responsible for the entire job, for the proper coordination of work, for the proper cooperation of workmen and the extent of work. (Subheadings have been incorporated only for convenience and no responsibility is claimed by the Architect or Owner for the extent, limits, or amounts of any subcontracts.)

12. EXPLANATIONS: All explanations to Contractor shall be agreed to in writing before signing of contract. No oral interpretations shall be valid.

13. SITE: Contractor shall visit site to ascertain full conditions under which work is to be done. Failure to do this will not relieve the Contractor of any responsibilities thus encountered.

14. DEBRIS, CLEANING: All trees, boulders, stumps, etc., within 10 feet of the building shall be removed. All debris shall be removed from the premises before final payments. At completion of job, building shall be left "Broom Clean" and all glass shall be washed and cleaned.

15. SPECIFICATIONS: This specification is intended to supplement the drawings and, therefore, it will not be its province to mention any part of the construction or materials which the drawings are competent to explain and any such omission shall not relieve the Contractor from carrying out such portions so indicated.

Should anything be omitted by both the drawings and the specifications which is necessary to a clear understanding of the work, or should an error appear in either of the various instruments furnished, it shall be the duty of the Contractor to notify the Architect.

Anything not expressly set forth in either the plans or specifications but which is reasonably implied shall be furnished and performed the same as items specifically mentioned.

16. OMISSIONS: The following will be supplied direct by the Owner and shall not be included in the contract.

1. All finish Grading and Seeding.
2. Sidewalks and Drives.
3. Shades and Venetian Blinds.
4. Range, Refrigerator, and Washer.
5. Screens and Storm Sash.
6.
7.
8.
9.
10.

17. SERVICE: All mechanical equipment and installations under this contract shall be guaranteed and serviced without charge for one year after date of acceptance.

18. ADDITIONAL ITEMS:

SITE WORK

1. PRELIMINARY LAYOUT: All property lines shall be established by the Owner. Building shall be located and staked out by the Contractor in the presence of the Owner. The Owner shall establish the main floor level, and the Contractor shall work accurately to all lines and grades so established.

2. CLEARING SITE: Remove from site of building all trees and obstructions within ten feet of the building and driveway. Any other trees on the property liable to damage during the construction shall be adequately protected.

3. EXCAVATION: Top soil at building location shall be removed and stored separately for final grading. Do all necessary excavation required for construction of basement, foundation, footings, etc. Excess material, not used in grading, shall be removed from property. Bottom of all excavation shall be level and solid.

4. GRADING: Backfill all excess excavation at the proper time. Any additional fill shall be provided at $ ______________ per yard. Spread all top soil over property and rough grade away from building to natural grades.

5. GRAVEL FILL: Where draintile is used, provide a minimum of 12 inches of coarse gravel before backfilling.

Where fill is required under slabs, same shall be porous material free of clay, loam or debris. No cinders permitted.

All fill to be solidly tamped and settled.

shall

6. DRIVEWAY: Contractor shall not cut present curb and replace, as per local requirements.

Contractor shall install driveway from property line to garage, of 6" compact gravel.

7. WALKS: 4-inch concrete walks shall be constructed where Owner directs at $ ____________________ per square foot.

CONCRETE AND MASONRY

1. CONCRETE WORK: Footings and trench-type walls – 2000# test.
   Floors and Walks – 3000# test.
   Where square-cut trenches are practical, no forms are necessary. Top of all work to be properly aligned, square, and straight. All joints of adjoining pours keyed and doweled with reinforcing rods.

2. FLOORS AND WALKS: To be straight and smooth and free from depressions.
   Where exposed, steel trowel.
   Where covered, wood float.
   Pitch where necessary for drainage.

3. BLOCK WORK: All subgrade – standard concrete block.
   All above grade – standard cinder, or slag, block.

   Block to be square and straight. Coursing to be level and true; bond all courses. End block, slabs, precast lintels, and standard miscellaneous shapes to be installed as required. Install reinforcing mesh every fourth course in all block work above grade.

4. CHIMNEY: Each unit to be lined with terra-cotta flue lining with a cleanout and thimble of cast iron.

5. FIREPLACE: Line with a minimum of 2" firebrick bedded in fireclay.
   Install cast-iron chain control damper.
   Hearth Material ____________________
   (tile, brick, or slate)
   Facing Material ____________________
   (brick, marble, or stone)
   Subhearth of masonry material 16" in front of fireplace.
   Where ash pit is indicated, provide cast-iron ash dump.

6. MORTAR: All Mason work to be laid in mortar cement of standard manufacture mixed according to manufacturers' instructions.
   Strike all joints.

7. DRAINTILE: Entire perimeter excavated sections: 4" porous crock; tape joints; 12" gravel fill over; connect to sewer, or dry well.

8. ANCHORS, ETC.: Where indicated and required.
   Plate Bolts – 3/8" by 6" – 4' O.C.
   Sleeper Clips – "Bull Dog" 16" O.C. each way in slabs.

9. SILLS: 2" Bluestone, full width of openings; one piece. Miter corners where necessary. (For block work only)

10. BRICK VENEER: Where so indicated on plans, provide brick anchored every fourth course, 16" O.C. with metal anchors. Brick to be $ ____________________ per M.

METALS

1. STRUCTURAL STEEL: As per plans, free from rust. Two coats lead and oil paint. Bolt sections as necessary.

2. FLASHING: 26-oz. galvanized iron at the intersection of roof and all vertical surfaces of chimney and dormers.

3. VALLEYS: Minimum 8" each side, 26-oz. galvanized iron.

4. GUTTERS: (See Details) Drain entire roof area. 26-oz. galvanized iron.

5. DOWNSPOUTS: 2" x 3" square section.

WOOD AND PLASTICS

1. GENERAL: This section to include all construction of interior partitions, walls, and roof, as shown on plans. Studs shall be doubled at all openings in interior and exterior walls. Increase header depth over all openings over 3'0" in bearing partitions and walls. All framing to be 2" clear of chimney construction, with double rafters where necessary. Place all necessary plastering grounds, and strip all exterior walls (Masonry).

2. MATERIAL: All framing and board lumber shall bear official grade mark and symbol and shall meet all the grading requirements of the Association in the grade.

All framing lumber shall be construction Douglas fir, or better.
All board lumber shall be no. 2 common yellow pine, or better.

Joist – Spans less than			
8'0"	– 2 x 6	16"	O.C.
" " " 12'0"	– 2 x 8	16"	O.C.
" " " 15'0"	– 2 x 10	16"	O.C.
" " " 18'0"	– 2 x 12	16"	O.C.

All lumber shall be dry and well seasoned, with moisture content not over 19 percent.

3. SHEATHING: 1" x 8" shiplap nailed diagonally, or insulating sheathing.

4. SUBFLOOR: 1" x 6" S4S nailed diagonally.

5. FURRING: Interior of all masonry walls to be plastered: 1" x 2" – 16" O.C.

6. ROOF SHINGLES: 235# Blue-black, class "C" Underwriters label – if asphalt shingles are used or 5X 16" Certigrade wood shingles, every fifth course doubled.

7. MILLWORK:
    A. EXTERIOR: Cornice: redwood or no. 1 white pine.
    Bevel Siding: 3/4" x 10" – 8 1/2" exposure – cedar.
    Vertical Siding: 3/4" x 12" and 1" x 2" battens.
    no. 2 white pine, or exterior grade Douglas fir plywood.
    Door Frames: 1 3/4" white pine rabbetted with 5 1/4" brick mould and oak sills.
    B. INTERIOR: Door Frames: 3/4" white pine, dadoed.
    Trim: Casing – 2 3/4" x 5/8" common white pine.
    Stop – 1 3/8" x 3/8" common white pine.
    Base – 2 1/4" x 5/8" common white pine.
    (No casing required for aluminum and steel sash)

All joints butted and coped, straight, set nails, sand smooth.

8. STAIRS: (A). Basement – (Where required) 1 5/8" plank treads.
3/4" board risers.
1 5/8" stringers.

(B). First Floor – (Where required) 7/8" housed stringer.
5/4" oak tread.
3/4" white pine riser.
Birch handrail.

9. CLOSETS: (<u>See Plans for Details</u>) All doors of 3/4" plywood on "Kennatrack" by J. McKenna Company, Elkhart, Indiana.
Clothes: Shoe and hat shelf and 3/4" pipe.
Linen: Five 14" shelves.

10. CAULKING: Caulk all joints at the intersection of wood, masonry and metal with "Tremco" caulking compound.

11. FINISH FLOOR: Under linoleum: 5/8" x 3 1/2" fir, end-match Tongue and Groove. All other floors – 3/4" x 2 1/4" #1 select red oak.

One layer of sound-deadening felt between sub and finish floor. All floors to be sanded smooth.

<u>THERMAL AND MOISTURE PROTECTION</u>

1. SUBGRADE WALLS: Waterproof exterior of all walls of all excavated sections with 1/2" cement-mortar plaster and one coat of asphaltic emulsion, as per manufacturers' directions.

2. SLABS: See plans for waterproofing.

3. INSULATION: Kimsul double-thick, aluminum-foil blanket, manufactured by Kimberly-Clark Co., to enclose entire habitable part of house.

<u>DOORS AND WINDOWS</u>

1. INTERIOR DOORS: 1 3/4" birch veneer, flush panel.

2. EXTERIOR DOORS: 1 3/4" stock. (See Elevations.) All doors weather-stripped.

3. WOOD SASH: Andersen casements, standard sizes.

4. ALUMINUM SASH: Reynolds aluminum casements, standard sizes.

5. STEEL SASH: Stock casements, standard sizes.

6. BASEMENT SASH: Detroit Steel Products Fenestra (sizes on plans).

7. GLAZING:

All sash and openings so indicated on plans shall be glazed in D.S.A. glass. All glass shall be properly set in nonhardening glazing compound suitable for sash material.

All mirrors shall be copperbacked, polished, 1/4" Plate Glass.

Double glazing (Picture Windows) shall be of standard manufacture of thickness to conform to manufacturers' standards.

FINISHES

CAUTION: Fill in all spaces, saying none if no finish is required.

1. SCHEDULE:

Basement –

	Wood Trim Material	Wood Trim Finish	Floor Material	Floor Finish	Sidewalls Plaster Finish	Ceiling Plaster Finish	Sidewalls Decorative Finish	Ceiling Decorative Finish
Laundry								
Coal Room								
Store Room								
Recreation Room								
Reception								
Dining Room								
Living Room								
Breakfast Room								
Kitchen								
Porch								
Stair Hall								
Stair								
Garage								
Bathroom No. 1								
Bathroom No. 2								
Bedroom No. 1								
Bedroom No. 2								
Bedroom No. 3								
Bedroom No. 4								

2. LATHING AND PLASTERING:

A. PLASTER BASE: "Gold Bond" Rocklath on floating fasteners.
Wire mesh at intersections of all surfaces.
Wire mesh at all lateral joints in ceiling.
Wire mesh at each second cross-joint in ceiling.
Corner beads on all exposed corners.

B. PLASTERING: Bath and kitchen – putty finish.
All other areas – fine sand finish, tinted.

Work to be three-coat method with brown and scratch combined.
Total thickness 5/8".

3. LINOLEUM:

KITCHEN: Floor, countertops and 16" backsplash.

BATH: Floor, walls, 4'6" high, and to ceiling at tub.

All floors to have 4" cove base, terminating with metal strip except where butted under cabinets.

All intersections of two surfaces to be coved. Owner selects all colors and patterns. Method of laying to be done in full accordance with manufacturers' recommendations and guarantees.

All terminating strips to be nonoxidizing metal or heat-proof plastic. Provide one piece sink rim.

4. TILE WORK:

NOTE: Work under this heading shall be verified with Owner.

ROOM ______________________ Floor______________ Sample No. __________
Walls______________ " " __________

ROOM ______________________ Floor______________ Sample No. __________
Walls______________ " " __________

All work shall be done in accordance with Specification of Tile Council of America.

5. PAINTING:

A. EXTERIOR: All masonry surfaces power sprayed with waterproofing, color as selected by Owner, as per manufacturers' specifications. Ten-year written guarantee to be furnished. Recommended products: "Thoroseal," Standard Dry Wall products, "Armor Coat," Armor Laboratories, Inc.

All wood surfaces, three coats of lead and oil paint.

All metal surfaces, one coat red lead and oil and two coats lead and oil paint.

B. INTERIOR: Entire uncovered plaster area of bath and kitchen, three coats lead and oil paint. All woodwork painted or stained. Any further work shall be verified in writing with the Owner. Wood floors, sealed, varnished, and waxed.

Colors to be selected by Owner. Paint to be of standard approved brands, delivered in sealed containers. Surfaces to be painted to be clean and free from dirt; knots and sap streaks to be sealed; putty all nail holes and cracks. All plaster abrasions and cracks to be cut out and filled smooth. Successive coats of paint shall not be applied until previous coat is dry.

SPECIALTIES

1. HARDWARE: Rough hardware (nails, bolts, etc.) by Contractor.
Finish hardware — allow $100.00.

2. KITCHEN CABINETS: 3/4" plywood sides, flush doors. Shelves of 3/4" redwood or white pine. Recessed toe space. Mortise all shelves to side and mortise drawer fronts.

3. ACCESSORIES: Allow $55.00 for medicine cabinet, towel bars, tumbler, and soap holders.

MECHANICAL

This item shall be checked with the Owner.

1. HEATING: All equipment shall be stock, installed as manufacturer directs. System to be guaranteed to heat entire dwelling to 72 degrees at __________ degrees outside temperature, and serviced free for the period of one year.

A. FIRING: (Check One) -

(a). ______________ Oil fired.
(b). ______________ Gas fired.
(c). ______________ Coal, hand-fired.
(d). ______________ Coal, stoker-fired.

B. SYSTEMS: (Select One) -

(a). _____ Forced Air – Duct work sheet metal 26-gauge, graduated trunk line. Adequate warm-air return registers in all rooms except bath and kitchen. Summer switch on fan.

(b). _____ Radiant – Complete system installed as per Bell & Gossett, using 100-degree circulating hot water. (Recommended for basementless one-story only). All piping "Byers" wrought iron, or Chase Company copper.

(c). _____ Hot Water – Complete system installed as per Bell & Gossett, using 160-degree circulating two-pipe hot water. All piping to be soldered joint hard copper.

C. CONTROLS: System to be completely automatic, thermostatically controlled. Water and stoker systems to have relay and low water controls. All electrical controls to be as manufactured by "Minneapolis-Honeywell."

D. FORCED AIR: If forced air is used, this Contractor shall provide T. C. crock cold-air returns for all rooms in unexcavated portions. All duct work in unheated spaces shall be insulated with 2" blanket.

E. FURNACE MAKE: ____________________ SIZE: ____________________

F. CONTROLS: ____________________

2. PLUMBING:

A. SYSTEM: Complete hot and cold water supply, sewer and drainage system to be installed with certified approval of the local jurisdictional authorities.

B. SEWERAGE: All underground work to five (5) feet beyond building, of extra heavy cast iron. Beyond building, glazed vitreous sewer pipe.

C. DISPOSAL: (a). City Sewer – Connect all sewage and waste water to municipal sewer.

(b). Septic Tank – Connect all sewage and waste to 550-gallon concrete septic tank and drain field, to be determined by soil conditions. No drain field closer than 100 feet from well.

D. CLEANOUTS: Full size of pipe.
At base of all vertical stands. At ends of each horizontal run. At each branch connection.

E. VENTS: Entire system to be vented with 4" vent through roof, properly flashed.

F. WATER SUPPLY: (a). Municipal Service: Install 1" copper service from main to building. Provide meter if required.

(b). Well: Casing size ______________________
Price per foot $ ______________________

Pump: To be determined by availability and depth of water. Minimum requirements – 400 gallons per hour at 40# pressure.

Pump Make ______________________

Pump Allowance $ ______________________

G. WATER PIPING: Soldered joint copper. Main runs 3/4" fixture taps 1/2". Hot and cold water to all required fixtures. Two outside hose connections with inside shutoffs and drains.

H. HOT WATER: (Select One)
(a). __________ Electric, 80-gallon insulated tank.
(b). __________ Oil, 30-gallon self-contained tank.
(c). __________ Gas, 30-gallon self-contained tank.
(d). __________ "Taco" exchanger with 30-gallon storage tank valved for twelve-month use directly from hot water boiler.

I. FIXTURES: The number, type, and manufacture of fixtures together with fittings, to be selected by the Owner and to become part of this specification.

	Fixture	Manufacture	Catalog No.	Fittings	Color
(a).					
(b).					
(c).					
(d).					
(e).					

ELECTRICAL

Entire electrical installation and equipment shall be in accordance with the National Electrical Code and the National Board of Fire Underwriters, and shall comply with the Local Ordinances and the Utility Company rules.

1. SERVICE: Three – #8 wire from meter and service panel to distribution panel in thin wall conduit.

2. BRANCH CIRCUITS: #12, two-wire Romex 660 watts maximum load per circuit.
   Motor Outlets – #12 Romex.

   Range – Three #8

3. SWITCHES: Bryant or Despard, single pole or three- or four-way, as per plans.

4. CONVENIENCE RECEPTACLE: Double flush type, Bryant or Despard.

5. PLATES: Plastic to match wall finish. Gang plates where required.

6. PANEL: Fuseless automatic circuit-breaker-type in pressed steel cabinet.

7. FIXTURES: Allowance – $150.00 Installation included in contract.

8. SIGNALLING: Entrance and service door, provide push buttons. Chimes, if selected, under fixture allowance.

9. ADDITIONAL ITEMS:

CHAPTER TEN

Construction-plan Reading

In Chapter 1, we asked the question: how does a mechanic learn to read drawings? In answer to that question, a list of procedures was set up. In Chapters 2 through 9, we learned these procedures in a step-by-step manner. Now we are ready to read a typical set of drawings for a commonly encountered type of structure.

The purpose of this chapter is to present a group of questions which will test our ability to read plans and, at the same time, constitute a most helpful review of the necessary visualization.

REVIEW QUESTIONS

Questions 1 through 65 are the type which will give us practice in knowing where to look, in a set of drawings and specs, for particular information. Many typical items found in drawings and specs are listed. For example, the first question mentions attic-floor joists. In our answer we simply state where we would look in the drawings and specifications for information about such joists. The second question mentions a terrace. In our answer, we simply state where information about that item can be found.

Readers are urged to answer all of the questions in writing and then check their answers with the correct answers shown in the Answer part of this chapter.

1. Attic-floor joists
2. Terrace
3. Garage

4. Ridge
5. Cornice
6. Distance between floors
7. Floor drains
8. Kinds of lumber
9. Height of window headers above floor levels
10. Room sizes
11. Downspouts
12. Picturelike views of windows
13. Scuttle
14. Saddle
15. Shingle symbol
16. Foundations
17. Floor-level height above grade
18. Stairs
19. Plumbing fixtures
20. Ash pits
21. Louvers
22. Chimney cap
23. Gutters
24. Kitchen-equipment arrangement
25. Dormers
26. Collar beams
27. Footings
28. Interior finish
29. Painting instructions
30. Horizontal dimensions
31. Shapes of roofs
32. Types of framing
33. Electrical layout
34. Kind of windows
35. Areaway floors
36. Pockets for sliding doors
37. Wall thicknesses
38. Floor materials
39. Window locations
40. Window sizes
41. Ceiling heights
42. Exterior wall finish
43. Foundation depth
44. Chimney flashing
45. Roof pitch
46. Plate heights
47. Door jambs
48. Headers
49. Insulation
50. Sheathing
51. Chimney flues
52. Soil-pipe location
53. Foundation key
54. Glass information
55. Furring
56. Moldings
57. Eave overhang
58. Floor framing
59. Wall framing
60. Joist size and spacing
61. Rafter ties
62. Size of ridge
63. Size of chimney
64. Hearth
65. Trimmer arch

DRAWING READING QUESTIONS

All of the followng questions are about Plates I through VI which accompany this book. On these plates, a typical modern house is represented by plan, elevation, section, and detail views. The drawings are exactly as the architect prepared them and as builders use them.

The following questions ask for information which is shown in the drawings. Read the questions and write answers to them. Check answers with the correct answers shown in the Answer part of this chapter.

66. What size front porch is required?
67. What type of floor material is required for the bathroom?
68. Where is linoleum required?
69. What kind of doors are required for the closets?
70. How many clothes closets are required?
71. Where is the fireplace located?
72. What size is the reception-room closet?
73. What are the overall dimensions of the house?
74. What kind of exterior walls are required?
75. How thick are the exterior walls?
76. How wide are the bedroom doors?
77. How many wall convenience outlets are required for bedroom 2?
78. Is access to the attic area to be provided?

79. What material is indicated for the fireplace?
80. What type finished flooring is required in the living room?
81. What size is bedroom 1?
82. How do we know where to locate the partition between bedrooms 2 and 3?
83. Where are glass partitions required?
84. What material is indicated for the hearth?
85. What material is required for the front and back porches?
86. How is the back porch roof supported?
87. What size is the dining room?
88. What plumbing equipment is indicated?
89. What size floor joists are required?
90. What size rafters are required?
91. What electrical equipment is indicated for the dining room?
92. How many windows are required for the living and dining room area?
93. What size flues are required?
94. How is the kitchen ceiling light controlled?
95. How many medicine cabinets are required?
96. What is the length and width of the living room?
97. What service-room equipment is indicated?
98. How far must the hearth area extend from the fireplace?
99. How many risers are required for the basement stairs?
100. Where is a wood railing required?
101. Where is the linen closet located?
102. How is the bathroom window located?
103. Are electrical outlets indicated for any of the closets?
104. Where is the soil pipe located?
105. What foundation material is required?
106. How thick are the foundations?
107. What are the porch foundations made of?
108. What are the pilasters made of?
109. Are steel beams required?
110. What size footing is required for the chimney?
111. What material is required for the basement floor?
112. What material is required for the areaways?
113. What size I-beams are required?
114. What type of flooring is required over the unexcavated area?
115. How is the chimney footing located?
116. How many electrical ceiling outlets are required for the recreation room?
117. How many wall outlets are required in the recreation room?
118. What kind of shingles is required?
119. What type exterior wall finish is required?
120. What kind of basement window is required?
121. What roof pitch is required?
122. Where are louvers required?
123. What are the porch railing and flower bed to be made of?
124. How far above the ridge is the chimney top?
125. What size windows are required for bedroom 1?
126. Where are fixed windows required?
127. What size footings are required?
128. How far below grade are the foundations, in connection with the unexcavated area, to extend?
129. How deep are the main foundations?
130. How far above grade is the floor level?
131. How deep are the porch foundations?
132. What is the ceiling height?
133. How wide is the chimney?
134. What material is the chimney made of?
135. How far above grade is the front-porch floor?
136. How thick is the concrete slab over the unexcavated area?
137. Where is metal roofing required?
138. How thick are the brick walls?
139. How far above grade are the brick walls to extend?
140. How far apart are the brick walls?
141. What is the chimney-cap material?
142. How much roof overhang is required?
143. What gutter material is required?
144. What kind of material is required for the flashing?
145. How wide is the fireplace opening?
146. What material is required for the fireplace facing?
147. What material is required for the fireplace trim?
148. What kind of glass is required for the glass partitions?
149. Is furring required in the kitchen?
150. How is the rough and finished floor to be laid on the concrete slab?
151. Is perimeter insulation required?
152. How thick is the required gravel bed?
153. Where is an expansion joint required?
154. What size draintile are required for the footings?
155. How is the foundation waterproofed?
156. How thick are the roof boards?
157. What size rake boards are required?

158. How thick is the sheathing?
159. What is the soffit made of?
160. What type and thickness of insulation is required?
161. What gutter depth is required?
162. How far above the basement floor must the bottoms of the floor joists be?
163. How is the foundation to be topped off?
164. What type interior wall finish is required?
165. What size plates are required?

ANSWERS

The following answers are for the questions set forth earlier in this chapter. The question numbers correspond to the answer numbers.

ANSWERS TO REVIEW QUESTIONS

1. On plan views.
2. On the first-floor plan.
3. On elevation, plan, section, and detail views.
4. On elevation and section views.
5. On elevation, section, and detail views.
6. On elevation and section views.
7. On basement views.
8. In the wood and plastics division of specs.
9. On elevation views.
10. On plan views.
11. On elevation views.
12. On elevation views.
13. On plan views.
14. Around chimney on elevation and plan views.
15. On elevation and section views.
16. On basement views.
17. On elevation views.
18. On plan, section, and detail views.
19. In connection with fireplace symbols on plan and section views.
20. On plan views.
21. On elevation and section views.
22. On elevation views.
23. On plan and detail views.
24. On elevation and section views.
25. On elevation and section views.
26. On section views.
27. On basement plans and section views.
28. On section and detail views.
29. In the finishes division of the specs.
30. On plan views.
31. On elevation views.
32. On section and detail views.
33. On plan views.
34. In the doors and windows division of the specs.
35. On plan and section views.
36. On detail views.
37. On plan and section views.
38. On plan and section views.
39. On plan views.
40. On elevation views or in schedules.
41. On elevation and detail views.
42. On elevation views and in specs.
43. On elevation or section views.
44. On elevation views.
45. On elevation and section views.
46. On elevation and section views.
47. On section views.
48. On section views.
49. On section views.
50. On section views.
51. On plan and section views.
52. On plan views.
53. On section views.
54. In the specs.
55. On section and detail views.
56. On detail views.
57. On elevation and section views.
58. On detail views.
59. On detail views.
60. On plan views.
61. On section views.
62. On section views.
63. On plan and section views.
64. On plan and detail views.
65. On plan and section views.

ANSWERS TO DRAWING READING QUESTIONS

Answers 66 through 104 are found in Plate I.

66. The size is 7′ 2″ by 7′ 4″.
67. Tile.

68. For the kitchen and service-room floors, for the kitchen cabinets, and for the lavatory top.
69. Sliding doors.
70. Five.
71. On the east wall of the living room.
72. The size is 4′ 0″ by 2′ 2″.
73. The over-all dimensions are 28′ 2″ by 55′ 0″.
74. Frame.
75. Six inches.
76. The width is 2′ 6″.
77. Three.
78. Yes, the scuttle at the west end of the hall is for that purpose.
79. Brick, firebrick, and T. C. linings.
80. Oak.
81. It is 11′ 6″ square.
82. The partition surface is 12′ 6″ from the interior surface of the west exterior wall.
83. In the reception room and in the bathroom.
84. Tile.
85. Concrete.
86. By a 4 by 4 post.
87. One dimension is to be 9′ 0″. By scaling, the other required dimension is found to be 8′ 6″.
88. A bathtub, two water closets, two lavatories, and a kitchen sink.
89. The joists are to be 2 by 10.
90. The rafters are to be 2 by 6.
91. One ceiling outlet, one wall switch, and two wall outlets.
92. Five.
93. The flues are 8 × 12 inches.
94. By two switches, one located near the dining-room door and one located near the service-room door.
95. Two.
96. The length is 19′ 6″ and the width, 12′ 0″
97. A washer and a dryer.
98. The distance is 1′ 6″.
99. Twelve.
100. In the service room along the stairs.
101. At the west end of the hall.
102. By the 3′ 2″ dimension.
103. No.
104. In the south wall of the closet for bedroom 1.

Answers 105 through 117 are found in Plate II.

105. Concrete block.
106. Eight inches.
107. Concrete.
108. Brick.
109. Yes, two.
110. Concrete.
111. The footing is to be 3′ 8″ by 5′ 8″ by 16″.
112. Corrugated iron.
113. The beams are to be 8 inches deep and weigh 17 pounds per lineal foot.
114. Concrete.
115. It is 5′ 6″ from the inside surface of the north foundation and 11′ 0″ from the inside surface of the west foundation.
116. Four.
117. Four.

Answers 118 through 132 are shown in Plate III.

118. Wood.
119. Wood siding.
120. Steel sash.
121. A slope of 5 inches in each 12 inches of run.
122. At gable ends.
123. Brick.
124. The chimney top is 3′ 0″ above the ridge.
125. Each sash is 32 by 24.
126. In the south wall of the living room.
127. The footing is 20 inches wide and 8 inches deep.
128. They should extend 2′ 8″ below grade.
129. The depth is 7′ 1½″.
130. The floor level is 1′ 3″ above grade.
131. The depth is 3′ 6″.
132. The height is 8′ 0″.

Answers 133 through 143 are shown in Plate IV.

133. The width is to be 4′ 8″.
134. Brick.
135. The front-porch floor surface is to be 8 inches above grade.
136. Six inches.
137. Over the front porch.
138. Eight and 4 inches respectively.
139. They are to extend 3′ 6″ above grade.
140. One foot.
141. Stone.
142. Sixteen inches.
143. Copper.

Answers 144 through 149 are shown in Plate V.

144. Copper.
145. The opening is 3′ 0″ wide.
146. Marble.
147. Birch.
148. Louvrex glass.
149. Yes, over the cabinets.

Answers 150 through 165 are shown in Plate VI.

150. By means of 2 by 2 sleepers on 16-inch centers.
151. No.
152. Six inches.
153. Between the edge of the concrete slab and the wall sill.
154. Four inch.
155. By a plaster coat ½ inch thick.
156. The boards are to be ¾ inch thick.
157. The rake boards must be 1 by 4s.
158. Sheathing must be ¾ inch thick.
159. Plywood.
160. Blanket-type, 2 inches thick.
161. The required depth is 3⅝ inches.
162. The distance must be 7′ 6½″
163. By solid concrete block.
164. Lath and plaster.
165. The plates must be 4 by 4.

Index

Index

House Plans

Plates I to VI

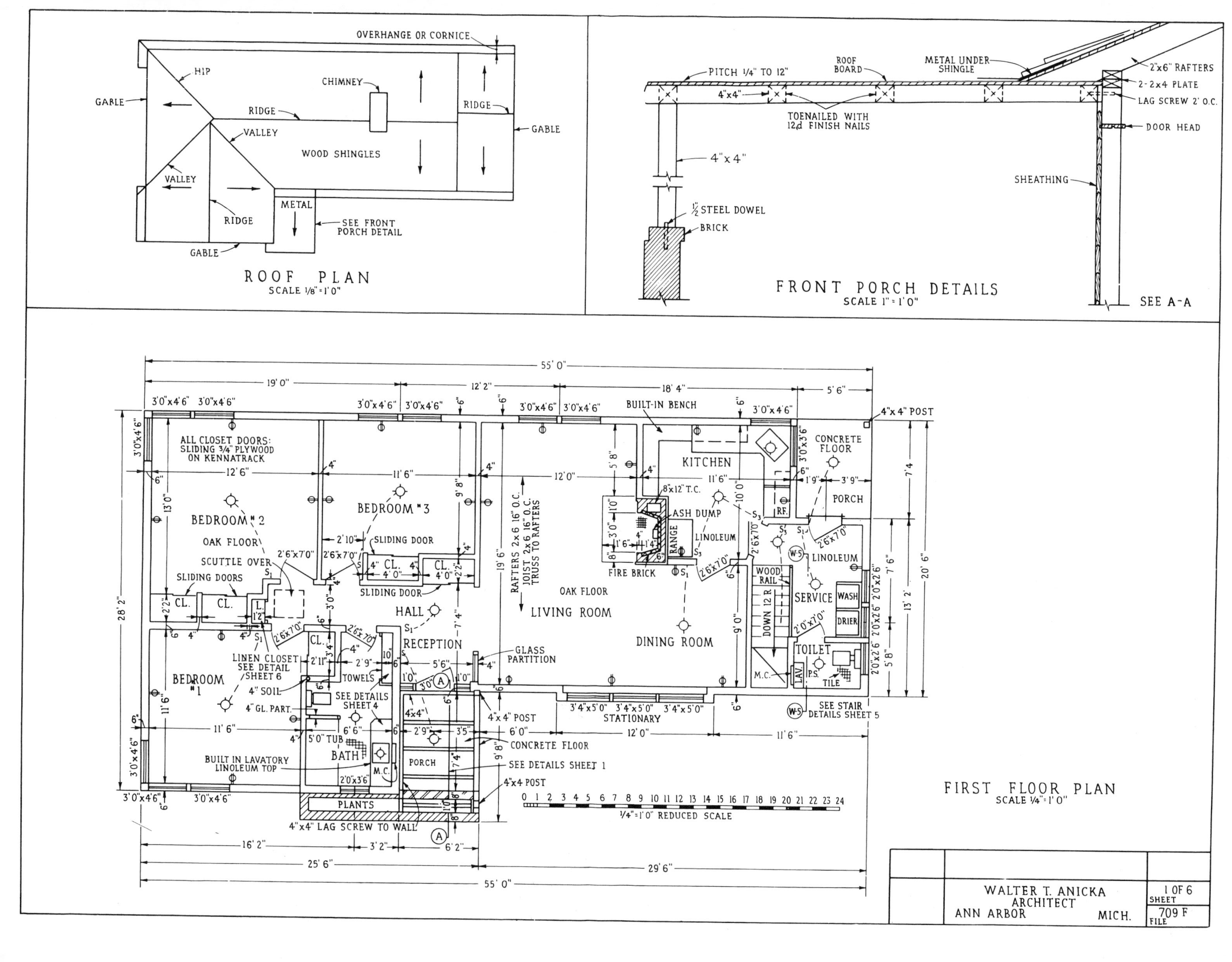

OVERHANGE OR CORNICE
HIP
CHIMNEY
GABLE
RIDGE
RIDGE
GABLE
VALLEY
WOOD SHINGLES
VALLEY
METAL
RIDGE
SEE FRONT PORCH DETAIL
GABLE
ROOF PLAN
SCALE 1/8" = 1' 0"
PITCH 1/4" TO 12"
ROOF BOARD
METAL UNDER SHINGLE
2"x6" RAFTERS
2-2x4 PLATE
LAG SCREW 2' O.C.
4"x4"
TOENAILED WITH 12d FINISH NAILS
DOOR HEAD
4"x4"
SHEATHING
1/2" STEEL DOWEL
BRICK
FRONT PORCH DETAILS
SCALE 1" = 1' 0"
SEE A-A
55' 0"
19' 0"
12' 2"
18' 4"
5' 6"
BUILT-IN BENCH
4"x4" POST
ALL CLOSET DOORS: SLIDING 3/4" PLYWOOD ON KENNATRACK
KITCHEN
CONCRETE FLOOR
PORCH
BEDROOM #2
BEDROOM #3
8"x12" T.C.
ASH DUMP
RANGE
LINOLEUM
OAK FLOOR
SLIDING DOOR
SCUTTLE OVER
SLIDING DOORS
CL.
FIRE BRICK
LINOLEUM
RAFTERS 2x6 16" O.C.
JOIST 2x6 16" O.C.
TRUSS TO RAFTERS
OAK FLOOR
LIVING ROOM
WOOD RAIL
DOWN 12 R
SERVICE
WASH
DRIER
SLIDING DOOR
HALL
RECEPTION
DINING ROOM
TOILET
GLASS PARTITION
LINEN CLOSET SEE DETAIL SHEET 6
TOWELS
BEDROOM #1
4" SOIL
SEE DETAILS SHEET 4
4" GL. PART.
4"x4" POST
STATIONARY
SEE STAIR DETAILS SHEET 5
5'0" TUB
BATH
CONCRETE FLOOR
BUILT IN LAVATORY LINOLEUM TOP
PORCH
SEE DETAILS SHEET 1
4"x4 POST
PLANTS
4"x4" LAG SCREW TO WALL
1/4" = 1' 0" REDUCED SCALE
FIRST FLOOR PLAN
SCALE 1/4" = 1' 0"
25' 6"
29' 6"
55' 0"
WALTER T. ANICKA
ARCHITECT
ANN ARBOR MICH.
1 OF 6
SHEET
709 F
FILE

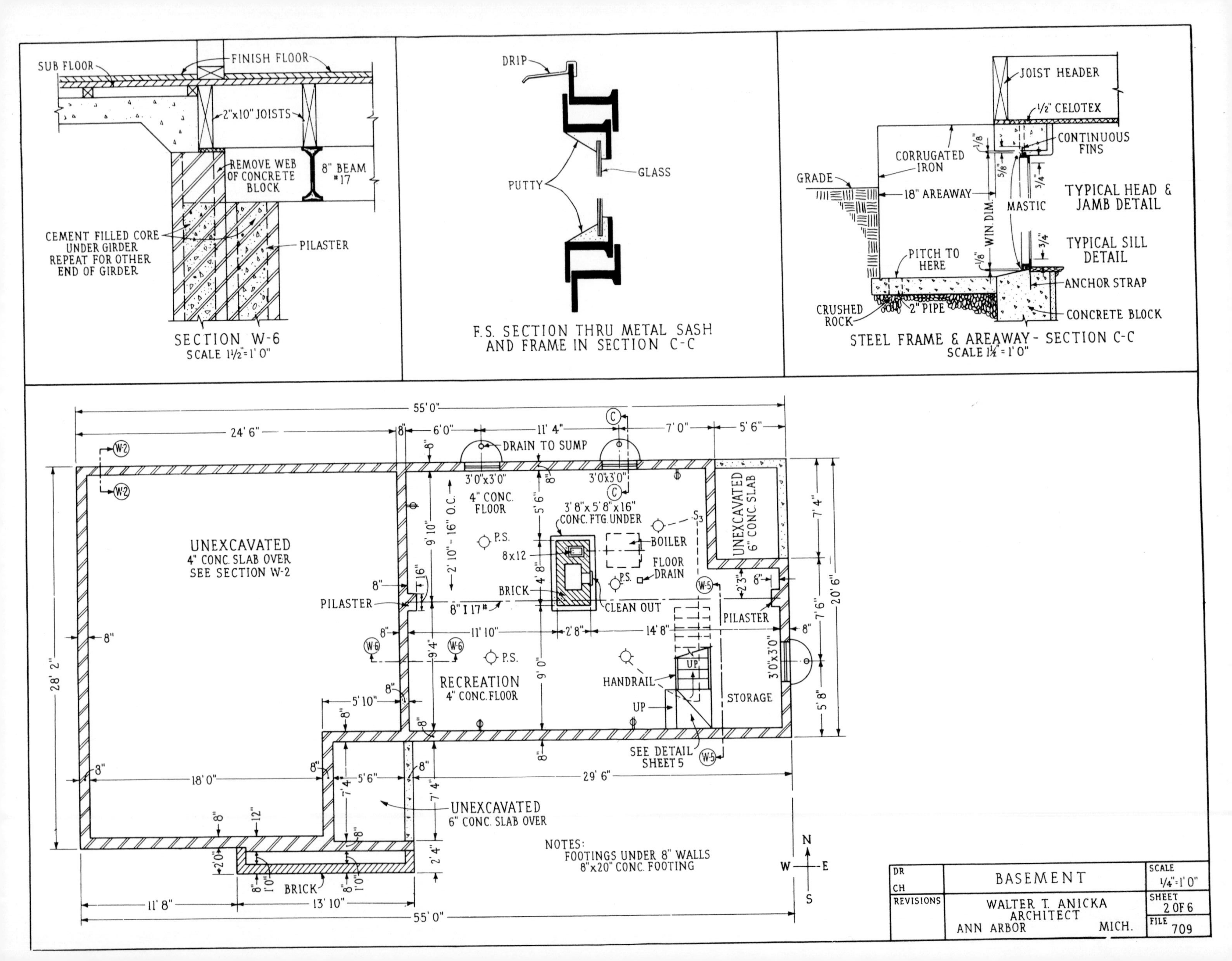
SUB FLOOR
FINISH FLOOR
2"x10" JOISTS
REMOVE WEB OF CONCRETE BLOCK
8" BEAM #17
CEMENT FILLED CORE UNDER GIRDER REPEAT FOR OTHER END OF GIRDER
PILASTER
SECTION W-6
SCALE 1½"=1' 0"
DRIP
PUTTY
GLASS
F.S. SECTION THRU METAL SASH AND FRAME IN SECTION C-C
JOIST HEADER
½" CELOTEX
CONTINUOUS FINS
CORRUGATED IRON
GRADE
18" AREAWAY
WIN. DIM.
MASTIC
TYPICAL HEAD & JAMB DETAIL
TYPICAL SILL DETAIL
PITCH TO HERE
ANCHOR STRAP
CRUSHED ROCK
2" PIPE
CONCRETE BLOCK
STEEL FRAME & AREAWAY - SECTION C-C
SCALE 1¼"=1' 0"
55' 0"
24' 6"
6' 0"
11' 4"
7' 0"
5' 6"
DRAIN TO SUMP
3'0"x3'0"
4" CONC. FLOOR
3' 8" x 5' 8" x 16" CONC. FTG. UNDER
BOILER
FLOOR DRAIN
8x12
BRICK
CLEAN OUT
P.S.
UNEXCAVATED 6" CONC. SLAB
UNEXCAVATED 4" CONC. SLAB OVER SEE SECTION W-2
PILASTER
8" I 17#
2' 10" - 16" O.C.
9' 10"
11' 10"
2' 8"
14' 8"
9' 4"
9' 0"
RECREATION 4" CONC. FLOOR
HANDRAIL
UP
STORAGE
SEE DETAIL SHEET 5
5' 10"
18' 0"
29' 6"
UNEXCAVATED 6" CONC. SLAB OVER
28' 2"
7' 4"
7' 6"
5' 8"
20' 6"
2' 4"
2' 3"
BRICK
11' 8"
13' 10"
NOTES:
FOOTINGS UNDER 8" WALLS 8"x20" CONC. FOOTING
N
W
E
S
DR
CH
REVISIONS
BASEMENT
WALTER T. ANICKA ARCHITECT
ANN ARBOR MICH.
SCALE ¼"=1' 0"
SHEET 2 OF 6
FILE 709

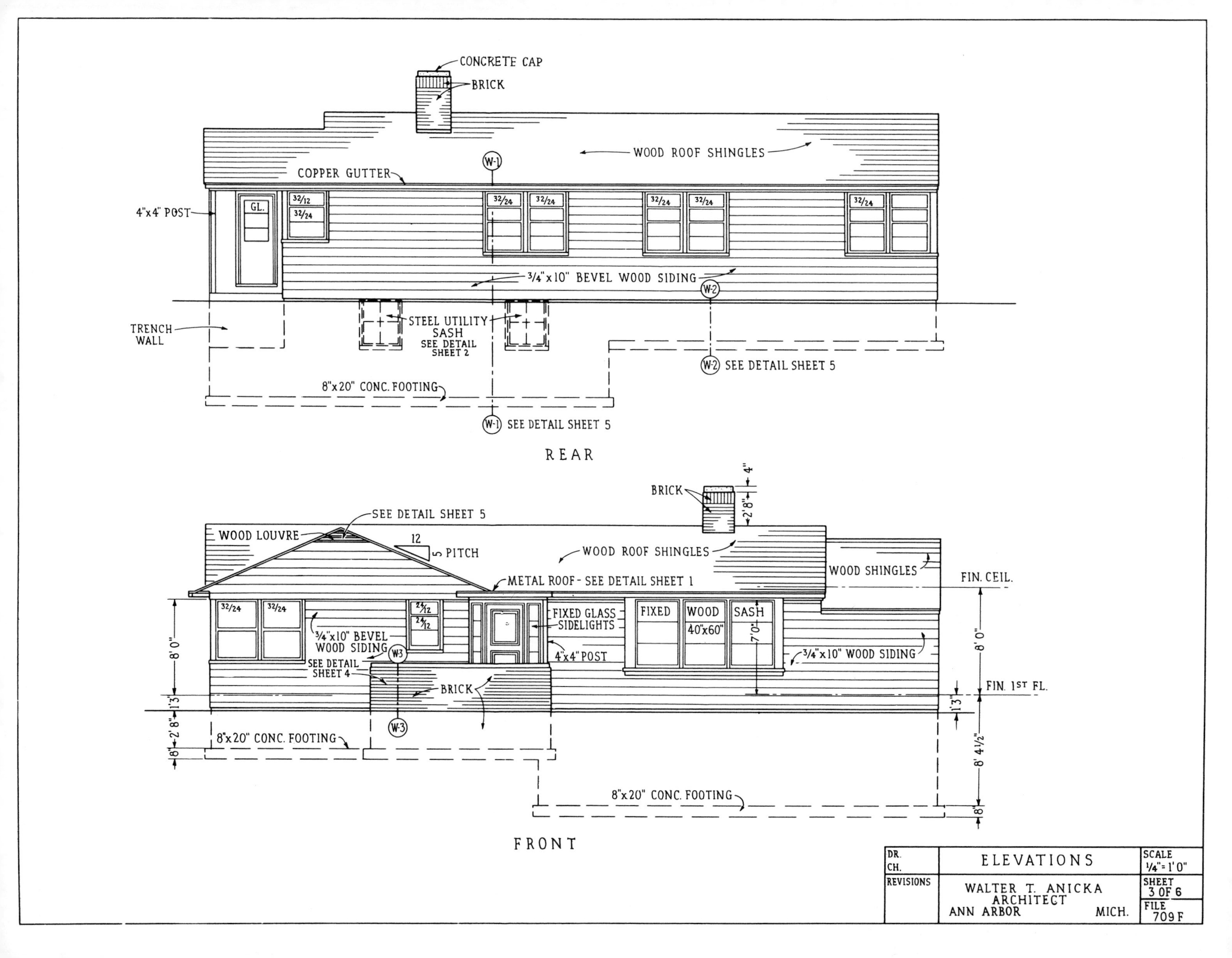
CONCRETE CAP
BRICK
WOOD ROOF SHINGLES
W-1
COPPER GUTTER
4"x4" POST
GL.
32/12
32/24
3/4"x10" BEVEL WOOD SIDING
W-2
TRENCH WALL
STEEL UTILITY SASH
SEE DETAIL SHEET 2
W-2 SEE DETAIL SHEET 5
8"x20" CONC. FOOTING
W-1 SEE DETAIL SHEET 5
REAR
BRICK
4"
2'8"
SEE DETAIL SHEET 5
WOOD LOUVRE
12
5 PITCH
WOOD ROOF SHINGLES
WOOD SHINGLES
METAL ROOF - SEE DETAIL SHEET 1
FIN. CEIL.
24/12
FIXED GLASS SIDELIGHTS
FIXED
WOOD
40"x60"
SASH
7'0"
3/4"x10" BEVEL WOOD SIDING
W3
SEE DETAIL SHEET 4
4"x4" POST
3/4"x10" WOOD SIDING
8'0"
BRICK
FIN. 1ST FL.
1'3"
2'8"
8"
8"x20" CONC. FOOTING
8' 4 1/2"
FRONT
DR.
CH.
REVISIONS
ELEVATIONS
WALTER T. ANICKA
ARCHITECT
ANN ARBOR MICH.
SCALE
1/4"=1'0"
SHEET
3 OF 6
FILE
709 F

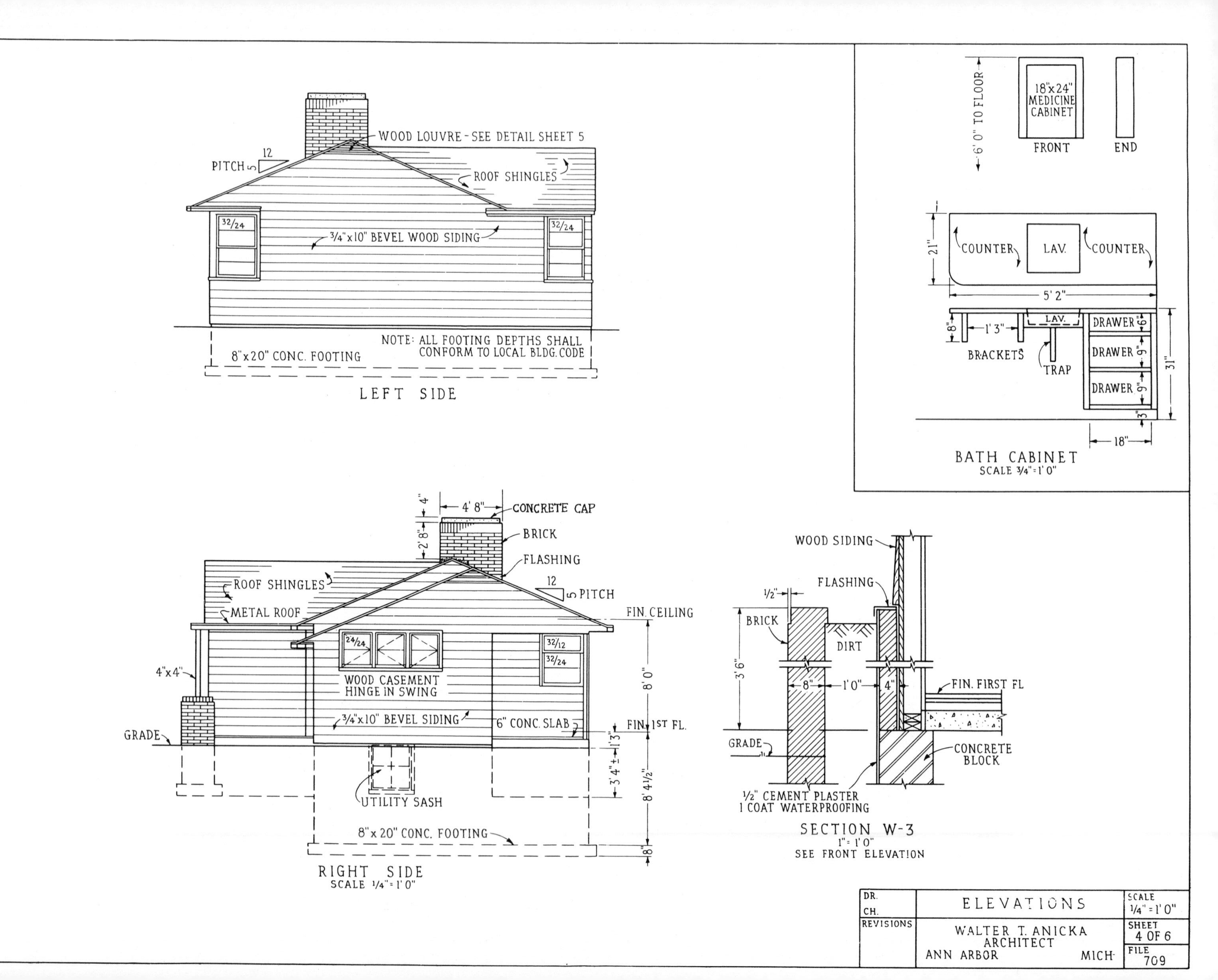
WOOD LOUVRE - SEE DETAIL SHEET 5
PITCH 5 12
ROOF SHINGLES
32/24
3/4"x10" BEVEL WOOD SIDING
NOTE: ALL FOOTING DEPTHS SHALL CONFORM TO LOCAL BLDG. CODE
8"x20" CONC. FOOTING
LEFT SIDE
6' 0" TO FLOOR
18"x24" MEDICINE CABINET
FRONT
END
21"
COUNTER
LAV.
COUNTER
5' 2"
LAV.
8"
1' 3"
BRACKETS
TRAP
DRAWER
DRAWER
DRAWER
6"
9"
9"
3"
31"
18"
BATH CABINET
SCALE 3/4"=1'0"
4"
4' 8"
CONCRETE CAP
2' 8"
BRICK
FLASHING
12
5 PITCH
ROOF SHINGLES
METAL ROOF
FIN. CEILING
24/24
32/12
32/24
4"x4"
WOOD CASEMENT HINGE IN SWING
8' 0"
3/4"x10" BEVEL SIDING
6" CONC. SLAB
FIN. 1ST FL.
GRADE
1'3"
3' 4" ±
8' 4 1/2"
UTILITY SASH
8" x 20" CONC. FOOTING
8"
RIGHT SIDE
SCALE 1/4"=1'0"
WOOD SIDING
FLASHING
1/2"
BRICK
DIRT
3' 6"
8"
1' 0"
4"
FIN. FIRST FL
GRADE
CONCRETE BLOCK
1/2" CEMENT PLASTER 1 COAT WATERPROOFING
SECTION W-3
1"=1'0"
SEE FRONT ELEVATION
DR.
CH.
REVISIONS
ELEVATIONS
WALTER T. ANICKA ARCHITECT
ANN ARBOR MICH.
SCALE 1/4"=1'0"
SHEET 4 OF 6
FILE 709

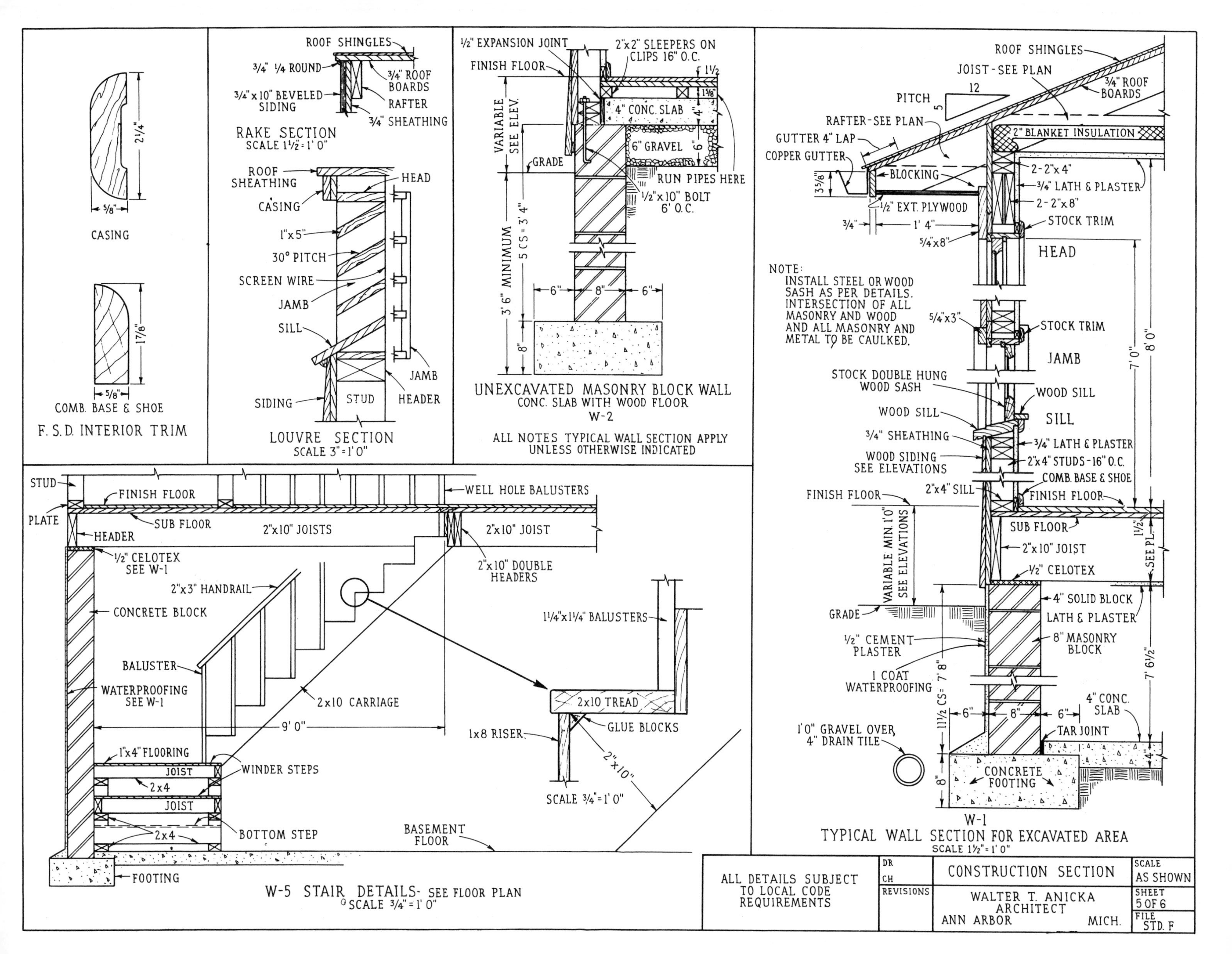

CASING
2¼"
⅝"
COMB. BASE & SHOE
1⅞"
F. S. D. INTERIOR TRIM
ROOF SHINGLES
¾" ¼ ROUND
¾" x 10" BEVELED SIDING
¾" ROOF BOARDS
RAFTER
¾" SHEATHING
RAKE SECTION
SCALE 1½" = 1' 0"
ROOF SHEATHING
CASING
1" x 5"
30° PITCH
SCREEN WIRE
JAMB
SILL
SIDING
HEAD
JAMB
HEADER
STUD
LOUVRE SECTION
SCALE 3" = 1' 0"
½" EXPANSION JOINT
FINISH FLOOR
2" x 2" SLEEPERS ON CLIPS 16" O.C.
4" CONC. SLAB
6" GRAVEL
RUN PIPES HERE
½" x 10" BOLT 6' O.C.
VARIABLE SEE ELEV.
GRADE
3' 6" MINIMUM
5 CS = 3' 4"
UNEXCAVATED MASONRY BLOCK WALL
CONC. SLAB WITH WOOD FLOOR
W-2
ALL NOTES TYPICAL WALL SECTION APPLY UNLESS OTHERWISE INDICATED
ROOF SHINGLES
JOIST - SEE PLAN
¾" ROOF BOARDS
PITCH
12
5
RAFTER - SEE PLAN
GUTTER 4" LAP
COPPER GUTTER
2" BLANKET INSULATION
BLOCKING
2 - 2" x 4"
¾" LATH & PLASTER
2 - 2" x 8"
½" EXT. PLYWOOD
STOCK TRIM
1' 4"
5/4" x 8"
HEAD
NOTE:
INSTALL STEEL OR WOOD SASH AS PER DETAILS. INTERSECTION OF ALL MASONRY AND WOOD AND ALL MASONRY AND METAL TO BE CAULKED.
5/4" x 3"
STOCK TRIM
JAMB
STOCK DOUBLE HUNG WOOD SASH
WOOD SILL
SILL
¾" SHEATHING
¾" LATH & PLASTER
WOOD SIDING SEE ELEVATIONS
2" x 4" STUDS - 16" O.C.
2" x 4" SILL
COMB. BASE & SHOE
FINISH FLOOR
8' 0"
7' 0"
SUB FLOOR
2" x 10" JOIST
½" CELOTEX
SEE PL.
VARIABLE MIN. 1' 0" SEE ELEVATIONS
GRADE
4" SOLID BLOCK
LATH & PLASTER
½" CEMENT PLASTER
8" MASONRY BLOCK
1 COAT WATERPROOFING
11½ CS = 7' 8"
7' 6½"
4" CONC. SLAB
TAR JOINT
1' 0" GRAVEL OVER 4" DRAIN TILE
CONCRETE FOOTING
W-1
TYPICAL WALL SECTION FOR EXCAVATED AREA
SCALE 1½" = 1' 0"
STUD
FINISH FLOOR
WELL HOLE BALUSTERS
PLATE
SUB FLOOR
2" x 10" JOISTS
2" x 10" JOIST
HEADER
½" CELOTEX SEE W-1
2" x 10" DOUBLE HEADERS
2" x 3" HANDRAIL
CONCRETE BLOCK
1¼" x 1¼" BALUSTERS
BALUSTER
WATERPROOFING SEE W-1
2 x 10 CARRIAGE
2 x 10 TREAD
GLUE BLOCKS
9' 0"
1 x 8 RISER
2" x 10"
1" x 4" FLOORING
JOIST
WINDER STEPS
2 x 4
SCALE ¾" = 1' 0"
BOTTOM STEP
BASEMENT FLOOR
FOOTING
W-5 STAIR DETAILS - SEE FLOOR PLAN
SCALE ¾" = 1' 0"
ALL DETAILS SUBJECT TO LOCAL CODE REQUIREMENTS
DR
CH
REVISIONS
CONSTRUCTION SECTION
WALTER T. ANICKA ARCHITECT
ANN ARBOR MICH.
SCALE AS SHOWN
SHEET 5 OF 6
FILE STD. F

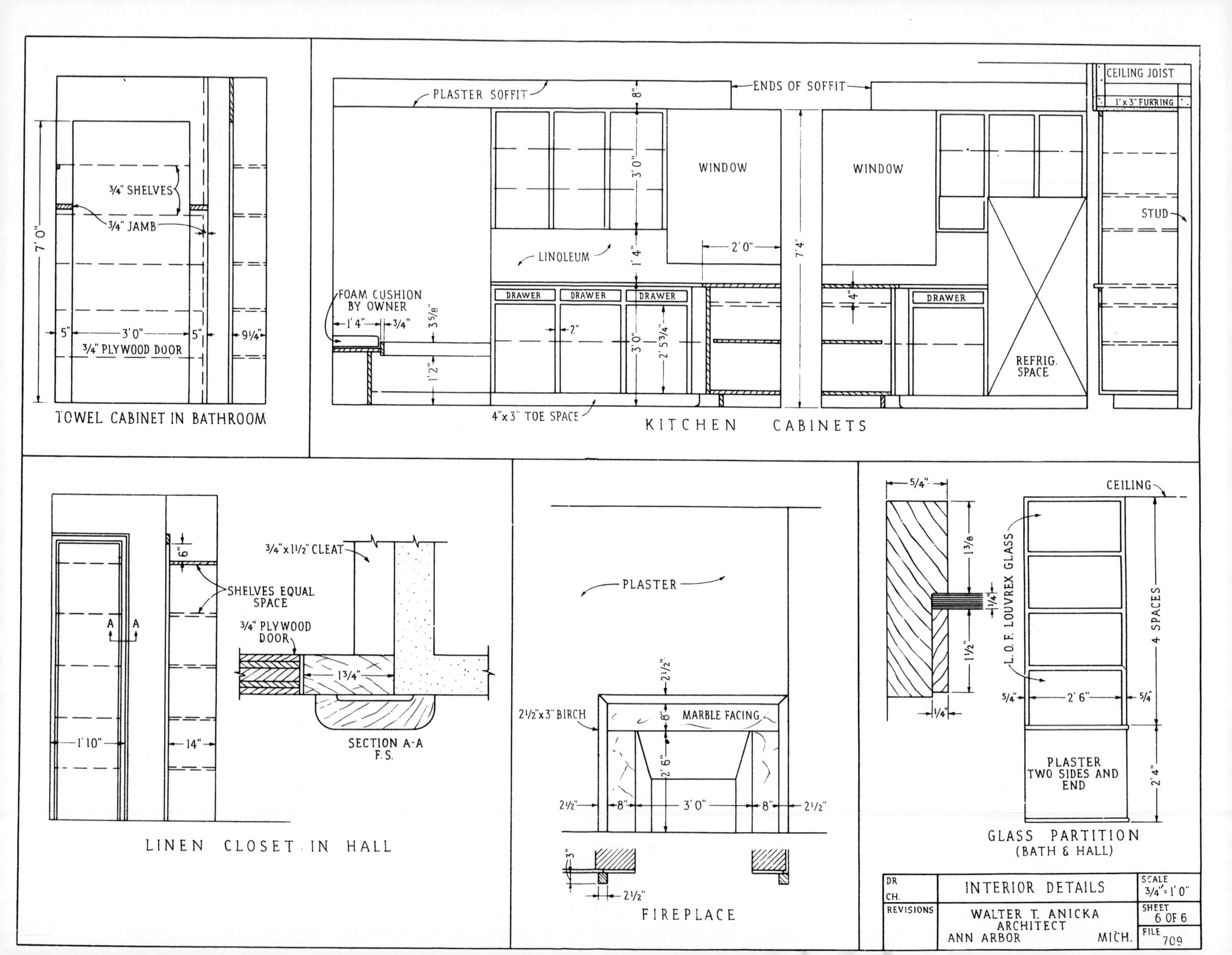

3/4" SHELVES
3/4" JAMB
7'0"
5"
3'0"
5"
9 1/4"
3/4" PLYWOOD DOOR
TOWEL CABINET IN BATHROOM
PLASTER SOFFIT
ENDS OF SOFFIT
CEILING JOIST
1"x3" FURRING
8"
3'0"
1'4"
WINDOW
WINDOW
2'0"
7'4"
STUD
LINOLEUM
FOAM CUSHION BY OWNER
1'4"
3/4"
3 5/8"
1'2"
DRAWER
DRAWER
DRAWER
DRAWER
2"
3'0"
2' 5 3/4"
4"
REFRIG. SPACE
4"x3" TOE SPACE
KITCHEN CABINETS
6"
SHELVES EQUAL SPACE
3/4"x1 1/2" CLEAT
3/4" PLYWOOD DOOR
1 3/4"
A
A
1'10"
14"
SECTION A-A
F.S.
LINEN CLOSET IN HALL
PLASTER
2 1/2"
2 1/2"x3" BIRCH
MARBLE FACING
8"
2'6"
2 1/2"
8"
3'0"
8"
2 1/2"
3"
2 1/2"
FIREPLACE
5/4"
1 3/8
1/4"
1 1/2"
1/4"
L.O.F. LOUVREX GLASS
CEILING
4 SPACES
5/4"
2'6"
5/4"
PLASTER TWO SIDES AND END
2'4"
GLASS PARTITION
(BATH & HALL)
DR
CH.
REVISIONS
INTERIOR DETAILS
WALTER T. ANICKA
ARCHITECT
ANN ARBOR MICH.
SCALE 3/4"=1'0"
SHEET 6 OF 6
FILE 709